Contents

About the author

Isobel Heathcote holds a B.Sc. from the University of Toronto and an M.S. and Ph.D. in physical limnology from Yale University. Her work experience is diverse, encompassing employment in consulting, government, and university teaching. In 1991, she joined the University of Guelph, where she is cross-appointed in Environmental Sciences and Environmental Engineering, and holds the position of Director, Institute for Environmental Policy. Her research interests center on integrated water management and watershed restoration, waste minimization in small industrial facilities, and environmental policy development.

About This Book

One of the greatest challenges in teaching environmental studies is that we as teachers cannot begin to imagine the problems our students will face five, ten or twenty years into the future. Although we have plenty of facts and figures on *past* problems, and how they were or were not resolved, there is no guarantee that the same solutions will work in a different place some time in the future. Each problem is a unique combination of biophysical factors and processes, and human cultures and systems. Often, it is the conjunction of the biophysical environment with human systems that creates an enviromental problem.

On graduation, environmental science students often find work in government agencies or consulting firms. In both cases, they must hit the ground running. On the first day, they may be confronted with a groundwater contamination problem, or a landfill selection issue, or some similar complex task. Through their environmental coursework, they will have received factual information about what groundwater is, where contaminated groundwater supplies exist, and so on, but they may have no idea how to attack a new problem that they have never seen before. Guided problem-solving in the classroom can give students the hands-on experience they need to feel comfortable tackling new problems.

This book therefore has several goals:

1. To present a number of *real environmental case studies* reflecting different cultures, values and biophysical environments around the world

2. To illustrate the *interaction of disciplines*—biology, chemistry, economics, sociology, engineering, agriculture, and so on—that lies at the heart of environmental management

3. To *demonstrate selected analytical techniques* that can help the environmental manager bring order to a complex problem

The cases in this book are intentionally "open-ended" in that there is no single "right" solution for any of them. Students are free to explore alternative approaches to each problem, and to use different analytical tools. Each case also offers an opportunity for more detailed discipline-specific analysis, such as population ecology, engineering design, or economic analysis, although this is not intended to be its primary use. Finally, this book offers an entry into a range of literature that draws from many countries and many disciplines, and which is therefore an excellent introduction to "real world" environmental management.

The Problem-Solving Framework

The problem-solving framework presented in this book consists of ten steps intended to clarify the goals and objectives of the decision makers, identify the mechanisms through which the problem operates, find a range of feasible solutions, and from that range choose "the best" in terms of the decision makers' goals. The following paragraphs summarize these ten steps. Each case study in this book then presents specific comments about how this framework can be applied to the case under study.

1. What is the problem?

The first and most important step in any environmental analysis is to determine the problem that we are trying to solve. This is often much more difficult than it might appear, simply because different stakeholders see the issue in from very different perspectives. Each case study in this book contains an analysis of "the problem", but students and instructors are certainly free to define "the problem" in different ways—indeed, some of the most interesting analysis may come from comparing the "best" solutions for different "problems". Some solutions may solve more than one problem, and may therefore be preferred over those that solve only a single, narrowly defined problem.

2. In what ways do human activities have impact on the natural environment to cause "a problem"? How do these mechanisms give you clues to possible solutions?

A large part of this step involves developing an understanding of the biophysical environment and how humans have had impact on that environment. This may include understanding physical, chemical, biological, sociological, economic, and other aspects of the situation. Understanding how a problem has developed can lead the analyst to an understanding of the pressures faced by stakeholders and the values placed on human activities and components of the environment. This in turn can provide useful insight to the kinds of solutions that may or may not work. Section 3 of each case study discusses the "environment" that is affected by the issue under study and gives some background on the biophysical environment and human activities impacting it.

3. What governments are responsible for the issue? Whose laws may apply?

Environmental managers tend to think of themselves as analysts, not lawyers. But the reality is that almost every solution will require approval from some government. And that means that almost every solution will have to comply with local, regional, state, and federal (possibly even international) laws. It's a good first step in any analysis to find out which governments will be granting such approval, and under which laws and policies. These rules may place important constraints on potential solutions and thus must be considered at the beginning, rather than the end, of the analysis.

4. Who has a stake in the problem? Who should be involved in making decisions?

In virtually every environmental problem many people and organizations have an interest, sometimes a financial interest, in how the problem is resolved. One of the most difficult challenges in environmental problem-solving is to decide who should be involved in making decisions. There's no single correct answer to this, although experience has generally shown that, over the long term, implementation is easier and more complete if a wide range of stakeholders have been involved in the decision making process, and if those involved have been given a meaningful role in decision making—not just in information dissemination.

5. *In the view of your decision-making group, what are the attributes of a satisfactory solution? In other words, when will you be satisfied that the problem is "solved"?*

It's not enough to identify the problem you are trying to solve—you also must decide what targets you are trying to meet. These targets can, and often should, be very specific in terms of numerical value, spatial extent and temporal application. If they are vague, decision making can deteriorate into time-wasting argument about what is and what isn't important, urgent, and so on. It's important to stress that there are no universal targets for environmental clean-up. Rather, targets are usually a local decision, reflecting local values and priorities, local land and resource use patterns, and similar factors. *It doesn't really matter what targets are chosen as long as they are clear and specific and the stakeholders agree among themselves.* Decision makers may also want to weight certain targets (cost is a common example) more highly than others (for example water quality); again, this is a community decision.

6. *How will you evaluate (test, compare) potential solutions?*

In almost every environmental problem, we will end up considering a number of potential solutions. It's easy to allow formal decision making to deteriorate into wrangling about whether this solution or that is "better". What do we mean by "better"? Well, our definition should relate clearly and specifically to the targets we set in (5) above—hence the need for clear targets. And in almost every case, we'll need some sort of analytical technique to decide objectively whether solution A meets our targets better than solution B. There are a variety of methods available for this type of analysis, such as from cost-benefit analysis, environmental assessment, statistical analysis techniques, computer simulation, community surveys, and expert interviews. Section 4 of each case study presents some relevant analytical methods; others are certainly available and feasible in most cases.

7. *What are all the feasible solutions to the problem?*

It's almost self-evident that we'll need to develop a set of possible solutions to the problem, and choose the best solution from among them. A long list of solutions can be developed simply by brainstorming, literature searches, and interviews with knowledgeable experts. Options that are not feasible can then be eliminated (for instance on the basis of cost or time constraints, or on their performance on environmental indicators like water quality). The decision makers can then focus their efforts on consideration of a short list of feasible solutions. The solutions can be compared with the techniques described in (6); sometimes, several techniques are used to give a range of feedback on each possible solution.

8. *Which solutions work "best" in terms of the attributes you identified in (5)?*

This step simply consists of reporting the results of analysis (steps 6 and 7) for the consideration of decision makers. If the targets and their relative importance (weights) are clear and the comparison of feasible solutions is thorough and objective, one or two solutions should emerge at this stage as "best". If several solutions are equally viable, the decision makers may want to consider secondary factors such as aesthetics in making their final decision.

9. *Which solution will be easiest to implement?*

Ease of implementation is an extremely important consideration in environmental problem solving. We can design the best remedial program in the world, but if we can't afford it or people won't accept it, it will simply never see the light of day. The factors that influence "implementability" are potentially diverse, including (but not limited to) cost, cultural and heritage values, fear of change, and failure to reach a workable consensus among stakeholders. Possibly the most common failure comes from inadequate involvement of stakeholders in decision making. Those excluded quickly can become opponents, and if their opposition is sufficiently vocal and powerful, it can create insurmountable obstacles to implementation simply by altering public (and therefore political) support for the plan.

10. ***What steps are needed for successful implementation? Who will pay? Who will monitor progress?***

This is the final step in problem solving: working out a step-by-step plan to find the necessary funds at the necessary time, to oversee implementation of the plan and monitor its impact, and to ensure that all laws and policies are met. It should follow logically from the previous steps, especially (3), (4), (8) and (9).

Using This Book

Practice using this problem solving framework should give students insight into the structure of a range of environmental issues and their commonalities and differences. Although this framework is by no means comprehensive or universal, it can give students confidence in approaching environmental issues that may be complex, controversial, and difficult to resolve—as most of them are.

Acknowledgements

I am indebted to many people who helped with the preparation of this book. In particular I would like to thank Erik Nielsen and Heidi McGregor, who compiled much of the research on which the book is based. Rob Ronconi's cheerful assistance, and his extensive contributions to Case Study 8, are also gratefully acknowledged. I owe many of the ideas and teaching approaches presented in this book to my undergraduate and graduate students, who are always my most lively and enthusiastic sounding board.

The International Council for Local Environmental Initiatives (ICLEI) and the International Marine Mammal Association were most generous in allowing access to their resources and in providing feedback on specific issues. My thanks also go to Dean McLaren, Environmental Coordinator, Town Planning Department, City of Wellington, New Zealand, and Elizabeth Hoenig of the City of Olympia, Washington's, Public Works Department, who provided reports on which two cases were based. Staff of the Harcourt Butler Technological Institute, Kanpur, India, the Danish Consulate, and the National Rivers Authority were also most generous with their time and materials. The Social Sciences and Humanities Research Council of Canada funded some of the research reflected in this book, and their support is gratefully acknowledged.

In a wide-ranging book such as this, errors in fact or interpretation are almost unavoidable. I am especially grateful for the assistance of J. E. de Steiguer, North Carolina State University; Chris Teplovs, University of Toronto; Gustavious P. Williams, Argonne National Laboratory; and James C. G. Walker, University of Michigan, who reviewed drafts of the manuscript and whose comments have helped make the book both more accurate and more useful. My thanks also to the staff at McGraw-Hill for their commitment to this project and their hard work in bringing it to fruition.

On a personal level, my husband Alan Belk and my children Elspeth Evans, Zoë Belk, and Edward Belk have provided ongoing support and forbearance even when the deadlines were tight and the author was irritable. My friend Mike Evans has been an enthusiastic supporter from the beginning, for which my thanks are also due. And finally my parents, Blake and Barbara Heathcote, have given me the unqualified love and encouragement without which no researcher, and no author, can succeed. My thanks to them all.

Isobel W. Heathcote
University of Guelph

Denmark

"Is wind power a viable alternative to conventional electric power generation?"

1. What Is the Background?

Denmark is a small country with no significant oil or hydroelectric resources, yet it is also a major consumer of electric power. How can such a country meet its energy demands efficiently?

In the energy crisis of the 1970s, Denmark was hit hard by two oil price increases. The country's government made a strategic decision at that time to exploit wind power as a major power source.

Today, Denmark's wind power system generates 70 watts of power per person—15 times the capacity of its next closest rival, the United States. Still, wind power makes up only 2.5 to 3% of Denmark's total power demand. The Danish government has set a target of 10% as the desired capacity for the year 2000.

The benefits of wind power are obvious: oil and gas savings, reduced pollutant emissions, low cost, and reduced susceptibility to international political forces. But there are also disadvantages to this method—considerations such as aesthetics, noise, inefficiency, and problems with reliability and accidental damage.

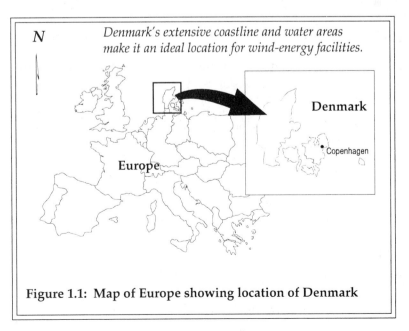

Denmark's extensive coastline and water areas make it an ideal location for wind-energy facilities.

Figure 1.1: Map of Europe showing location of Denmark

Storage is one of the main issues with this method of power generation. Consumption of electricity does not necessarily coincide with the presence of favorable wind conditions. So far, there is no good technology available for long-term storage of energy, so Denmark—and other countries—will always need another source of electric power.

What lessons can be learned from the Danish experience? Other countries, notably the United States and the United Kingdom, have access to the same technologies. What has allowed Denmark to move ahead so successfully in developing its wind power capacity?

? 2. What Problem Are We Trying to Solve?

Conventional approaches to the generation of electricity rely on fossil fuel resources (coal, oil), nuclear processes, or the force of falling water. Yet these resources are largely nonrenewable: fossil fuel and uranium reserves are finite, as are the number of good hydroelectric sites. Even countries rich in these resources can therefore benefit from development of a renewable energy program.

The energy potential of wind is enormous. Some estimates suggest that wind energy could supply 5 times the amount of electricity presently used worldwide. Smaller, less energy-consumptive countries could supply at least 25% of their energy demands from wind; in theory, up to 100% of demand could be met from this source. Why then has wind power been slow to gain acceptance?

Our problem in this case is to determine what factors have allowed Denmark, and some other European countries, to move ahead quickly on the development of wind power as an alternative to conventional generation technologies. In other words, what factors contribute to a country's ability to restructure its energy strategy?

As part of this analysis, we must also understand the advantages and disadvantages of wind power as an alternative to conventional power generation. These forces can be very important in driving a country toward or away from alternatives.

We can categorize these issues under a few main headings as follows:

1. *Environmental implications*

2. *Cost*

3. *Efficiency*

4. *Durability*

5. *Generation capacity*

We will examine each of these issues in the following section.

Table 1.1: European commitments to wind power	
Country	*Installed capacity*
Denmark	*1,500 MW by 2005*
Netherlands	*1,000 MW by 2000*
Germany:	
Schleswig Holstein	*1,000 MW by 2000*
Lower Saxony	*1,000 MW by 2005*
Italy	*300 MW by 2000*
Spain	*180 MW by 2000*
United Kingdom	*1,000 MW by 2000 (all renewables)*
European Community	*8,000 MW by 2005*
Source: World Watch Institute, 1993	

3. What Components of the Environment Are Affected, and How?

A brief introduction to wind power

For centuries, wind has been used as a source of power for transportation, water extraction and milling of grain. How is wind power used to generate electricity? The principle is similar to that used in other power generation technologies. The potential energy in wind is transferred to the kinetic energy of a turning blade, or turbine, which in turn induces the flow of electricity (see box).

In thermal electricity generation, fossil fuel is burned to create steam, and the pressure of steam then drives the turbine. In hydroelectric power generation, the turbine is turned by falling water. Tiny turbines—less than 2.5 cm in diameter—are even used to generate power for shop machinery.

Turbines are typically of two types: impulse turbines, driven by the force of a fast-moving fluid such as a jet of water, and reaction turbines, in which the rotor (the rotating part of the turbine) turns because of the weight or pressure of a fluid on a blade or plate. Old-fashioned water wheels are one type of reaction turbine. Hydroelectric power generation often employs impulse turbines, using a combination of a long pipe (penstock) and nozzles to increase water velocity and aim the water jet at the rotor.

Wind turbines are typically reaction turbines, and can be oriented around a vertical axis or a horizontal axis. Modern horizontal-axis wind turbines usually have two or three slender propellor-like blades mounted on a tower tall enough to catch the wind. Wind turbines installed in recent Danish developments employ three-bladed, horizontal-axis machines of this type, mounted on 22-m towers set in reinforced concrete bases.

> *There are two types of turbine: impulse turbines, driven by a jet of fast-moving fluid, and reaction turbines, pushed continuously by the pressure of water or wind.*

In a sense, these windmills are the modern evolution of the familiar Dutch and American farm windmills. Modern wind turbines adjust to the speed and direction of the wind, including changing the angle of the blades to allow the turbine to operate at a relatively constant speed.

Vertical-axis wind turbines have been around since the 1920s but are far less common than horizontal-axis turbines in modern wind-farm developments. Vertical-axis turbines have curved vertical blades positioned around a vertical shaft, rather like an upright egg beater. The vertical position of the blades allows the turbine to catch wind coming from any direction.

Wind turbines function best—are most efficient—in areas that have the highest wind energy and the fewest obstructions. The Danish government has classified the landscape of that country into four types, defined primarily by their roughness:

Class 0: Water areas, including ocean areas, that have the highest relative energy
Class 1: Open country areas with few bushes, trees, and buildings
Class 2: Farmland with scattered buildings and hedges, with land divisions occurring at intervals greater than 2 km
Class 3: Built-up areas, forests, and farmland with many hedges

Careful attention to siting of wind turbines and wind farms (groups of turbines) is therefore an important consideration in optimizing generation potential and perhaps minimizing the risks of accident or interference from other land-based activities. In Denmark, there is increasing interest in siting wind farms offshore, in areas of highest energy (and thus highest power generation potential) and least impact on residential areas. The more viable, efficient, and reliable a technology is seen to be, the more it will gain acceptance from the community and from the government. The following factors are important in this analysis.

The environmental implications of wind power

Although it clearly has a number of environmental benefits—less polluting, less use of scarce natural resources—wind power also has several important negative impacts on the environment.

Impacts on birds

Bird impacts are of two kinds. First, birds can be killed or injured in collisions with wind turbine blades or towers. Research has shown that these collisions are more frequent and more serious with wind turbines over 50 m in height. Particular concern has been raised about the construction of wind farms in offshore ocean waters, which may be major migration routes for birds. Many areas, whether coastal or inland, experience periodical, often seasonal, high densities of birds. Locating wind farms in these areas has in the past provoked considerable public protest, especially from wildlife protection organizations. Part of the debate around wind turbine impacts on birds relates to the inadequate data base available for evaluating these effects. More research is needed to develop better bird-safe turbine designs and better approaches to siting them.

The second type of bird impact relates to the noise of the turbine and its actual or potential effects on nearby breeding or resting birds. Research on this latter impact is even scarcer, so firm conclusions will have to wait until the problem has been studied in more depth.

Box 1.1: Generating electricity with a turbine

A generator is a device that converts mechanical energy to electrical energy. Generators produce energy by the principle of electromagnetism, a concept that was described by Michael Faraday and Joseph Henry in the 1830s. In laboratory experiments, these scientists discovered that it was possible to induce the movement of electrons—electricity—in a coil of copper wire by moving the coil near a magnet or the magnet near the coil.

A simple generator can be constructed out of a loop of wire and a U-shaped magnet. Rotating the loop of wire between the poles of the magnet interrupts the magnetic fields of the poles and induces electricity in the loop. It works this way: one side of the loop of wire cuts upward through the magnetic field, while the other side cuts downward. Halfway through the loop, the forces are neutral. On the bottom half of the loop, the reverse occurs.

Electricity that flows one way in the top half of the loop of wire thus flows in the opposite direction in the bottom half of the loop. This process is referred to as "alternating current" because the current, or voltage, travels in one direction half the time (top half of the loop) and in the other direction the other half of the time (bottom half of the loop). One complete loop is called a cycle; the number of cycles in a second is therefore the frequency of the voltage or current. Frequency is measured in units called "hertz", where one hertz is equal to one cycle per second.

Noise

As the rotor of a wind turbine turns, it creates two types of noise. The first type is mechanical in origin, for instance the noise made by the spinning rotor and by vibration of the machinery itself. Research has shown that this type of noise is most problematic in smaller turbines with rotor (blade) diameters less than 20 m. The second type of noise, more apparent in larger rotors (over 20 m in diameter) is aerodynamic in origin and results from the passage of wind over the blade structure. This type of noise is familiar to us as the howling or singing of wind over cables and similar structures.

Many countries have established allowable noise levels. Often, such limits are established based on baseline wind speed (in the case of the Netherlands, that baseline is 5 to 7 m/s) because perception of noise, and the presence of background noise, will vary depending on prevailing wind speed.

When an acceptable level of noise emission has been established (40 decibels at windspeeds of 5 to 7 m/s in the Netherlands), planning authorities can make reasoned decisions about where residential areas and other land uses can be located without noise impact. Where there are a number of wind turbines operating together (a wind farm), additional buffer distance may have to established (or wind turbines spaced farther apart) because of the potential for noise to be generated by interactions among the turbines.

Aesthetics

The appearance of wind turbines has long been citied as a major public concern about this technology. Clearly such an issue is of most concern where population density is highest. Densely populated areas are least appropriate for such technology in any case because of reduced wind potential, so this concern may in fact be less important than it seems at first glance.

Nevertheless, aesthetics have played an important part in the design of modern wind turbines, and further advances in this area are being made all the time.

Safety

Although safety issues have not been the focus of much research, wind turbines have the potential to impact on the safety of people working in, living in, or simply passing through a wind farm area. Such a risk might arise from blade fracture (materials failure), from structural damage (for instance by airplane traffic or bird impact), or from structural failure (toppling of the tower structure under extremely high winds).

What is an "acceptable risk" to safety? The decision will vary with the jurisdiction and the values and priorities of its constituents. In the Netherlands, a political decision was made that an "acceptable risk" of accident is equal to or less than 1 in 10 million. If we are speaking of accident from wind turbine operation, this means that estimated risk of death from a wind

---

Box 1.2: What is an acceptable risk?

Different groups of people have different ideas about what level of risk is "acceptable". Their perception may be affected by the other voluntary or involuntary risks to which they are exposed. Voluntary risks include activities like smoking, white-water canoeing, and rock climbing that have clear health and safety risks. Involuntary risks include consuming food and water contaminated with toxic substances, living in densely populated areas, and so on. Perception of "acceptable risk" is often closely tied to perception of what risks are already accepted by the population and how the proposed activity compares to accepted, everyday activities in terms of risk.

Research has shown that people tend to be very apprehensive about the prospect of unfamiliar technology like wind power. When wind turbines are built, however, public acceptance tends to increase rapidly: the reality isn't as bad as was feared.

What do we conclude from this? One analysis notes that apprehension and then rapid acceptance may be a result of inadequate knowledge of the technology in the general population. While acceptance of in-place turbines is encouraging, some analysts believe that it cannot be stable if the population isn't well informed. A serious wind turbine accident, for example, might instantly reverse public support to public opposition— and throw a country's wind power strategy into disarray.

Public education about wind power may therefore be an essential element in an effective national strategy to promote alternative energy sources and clean technologies.

turbine accident must be less than 1 in 10 million for that wind turbine installation to be considered acceptable in that society.

Little is known about the rates and risks of such occurrences. More complete information, and perhaps computer simulation of wind turbine performance, might form the basis for land use planning decisions: where houses and roads can be built safely, and where buffer zones need to be established.

Efficiency and durability

In recent years, the development of technology has allowed huge increases in the generating capacity of wind turbines. The latest turbines are now capable of generating 500 kW, up from an average capacity of 55 kW in 1980. This increase is a result of improved rotor design in modern turbines, allowing better wind capture and initiation of turning at lower wind speeds. Modern turbines can therefore generate electricity over a longer time period (and at lower wind speeds) than was possible with earlier technology.

A major challenge in the design of wind turbines has been the problem of how to keep the rotor turning at a constant speed (usually set to the cycle speed of the circuit) despite varying wind speeds. Getting the rotor turning at low speeds is a challenge that has been met successfully in modern designs. More difficult has been the development of brakes so that the rotor does not spin out of control at high wind speeds or during gusty periods. Early air brakes were subject to frequent failure, probably the most common cause of failure in early turbine models. Newer aerodynamic brakes come on slowly as wind speed increases and function with a much lower failure rate.

Cost

Two main components: capital costs and operating/maintenance costs

As modern wind turbine technology is refined and improved, the costs of turbine operation have dropped significantly. Estimated capital (start-up) costs are probably on the order of $1,000 U.S. per kilowatt of generation capacity, based on installations in the United Kingdom and elsewhere in Europe. Operation and maintenance costs vary considerably, depending on the jurisdiction where the turbine is installed, because of differences in the costs of labor and equipment rental. These costs are typically highest in the first year of operation. In 1989, the average costs of operation in the first year were estimated at about $0.006 U.S. per kilowatt-hour based on a variety of European and American installations. As the turbine ages, operation and maintenance costs drop significantly.

How long will the structure last?

To estimate the total "cost" of an installation, it is necessary to estimate how long the facility will last. A 25-year lifespan is often used for major capital works of this type, but the high failure rate in the first year or so of operation has led some U.S. manufacturers to recommend a shorter planning period, perhaps 12 years or so. Other manufacturers recommend planning for the 25-year period, but with the understanding that the facility will be overhauled and refurbished halfway through its life. Some analysts suggest that the savings in annual operation and maintenance after the first year are likely to be balanced by mid-life refurbishment costs, so an accurate cost can be estimated as the capital cost plus an ongoing operation-plus-refurbishment cost of 0.6 cents per kilowatt-hour per year over a 25-year planning period.

Table 1.2: Land used by major electricity-generating technologies in the United States	
(square meters of land per 1,000 megawatt-hours over 30 years)	
Technology	**Land occupied**
Coal (incl. coal mining)	3,642
Solar Thermal	3,561
Photovoltaics	3,237
Wind (turbines and service roads)	1,335
Geothermal	404

Source: World Watch Institute

Dismantling and decommissioning costs

What about the costs of dismantling the facility once its useful life is over? These costs are called decommissioning costs. It is difficult to know with certainty how large those costs will be 25 years into the future, but a reasonable estimate might be that the costs of building the facility will approximately equal the costs of decommissioning it.

Overall economic viability?

The overall profitability of the installation can therefore be roughly estimated as the annual revenues expected from the sale of electricity (its "value" to the power system), less the annual costs of operation and maintenance. Some estimates suggest that this difference is about $0.015 per kilowatt-hour. If this estimate is correct, wind power's claim to be a highly cost-effective would seem to be well substantiated. Perhaps more important from an implementation perspective is the conclusion that wind power is a commercially viable proposition, and one worthy of government support.

4. How Can I Analyze This Information?

In the early 1980s, California far outstripped the rest of the world in the production of wind energy. Today, however, Denmark, Germany, the Netherlands and the United Kingdom are rapidly increasing their use of this technology, while U.S. development of wind farms has slowed considerably. It is estimated that by the end of 1996 Europe will be out-producing the United States in wind power, not just on a per-capita basis but in total kilowatt-hours.

What factors have allowed this rapid adoption of wind power to occur? What lessons can be learned for other jurisdictions? An analysis of these questions cannot be undertaken with quantitative tools such as statistics or cost-benefit analysis. Rather, it is a case of understanding the underlying forces that drive major public decisions.

In the case of wind power in Denmark, we can suggest the following logical sequence of forces:

1. Denmark, like some other European countries, lacks domestic reserves of fossil fuel and uranium for thermal power generation and lacks adequate sites for hydroelectric power generation.

2. As a result, Denmark must rely on other countries to supply the raw materials for generation of electricity.

3. Reliance on these other countries makes Denmark vulnerable to political forces and price changes that are beyond its control.

4.   Therefore it makes sense to develop a national strategy for alternative energy sources.

5.   Having made a national commitment to alternative energy sources, Denmark should first develop those sources that are most cost-effective—that deliver the greatest capacity for the lowest cost.

6.   Wind power has been proved to be an efficient and inexpensive technology, and Denmark has access to excellent wind-generation sites, especially in offshore regions. Therefore wind power should be developed first.

Other sequences are possible. We could begin with the assumption that coal-, gas-, or oil-fired thermal power generation is costly and polluting, and therefore a cleaner technology should be found. Wind power is clean and cheap and therefore a good choice.

Or we could say that the European Union's commitment to 8,000 megawatts of wind power by 2005 requires Denmark to pull its weight and proceed with installation of wind power capacity.

There are elements of truth in all these sequences, but the sequences also share major assumptions:

1.   That sufficient political will must be present to encourage the development of wind power capacity, including practical measures such as clearly articulated policy goals, appropriate legal frameworks that favor wind power, and economic assistance in the form of subsidies for groups and individuals wishing to develop that capacity.

2.   To gain political acceptance, a technology must make sense for the society in which it will operate. In this case, it must be seen as socially and economically at least equivalent to, if not better than, conventional technologies.

3.   Political acceptance is not enough. There must also be support from the public, from research organizations, and from industry. This support should reflect broad societal needs to improve the efficiency of the technology and reduce its impact, and can range from research and development through installation and operation stages of the technology.

> **Box 1.3: Choosing a planning period**
>
> *Over what period of time should you plan for major capital works like a power generation facility? The choice can have important implications for assessing the cost and feasibility of a major project.*
>
> *The choice of a planning period is ultimately a personal one. In making the choice, the analyst should consider factors such as the probable lifespan of materials used in construction and their likely replacement frequency, changing land use patterns and a desirable interval for re-evaluation of major facilities, and planning periods typical for other major projects in the area.*

The story of wind power in Denmark is an illustration of the ways in which external political and economic forces can drive internal policy and technological change in a country. Many factors can initiate that change, but it requires a combination of government, industry, research, and public effort to maintain momentum and achieve national commitments to technological change.

5. How Can I Use My Findings to Reach a Solution?

Use the decision-making framework described in the Introduction to organize your thinking on this problem, as follows:

1. *What is the problem?*

In Section 2, we identified the problem as determining what factors contribute to a country's ability to restructure its energy strategy.

2. *In what ways do human activities have impact on the natural environment to cause "a problem"? How do these mechanisms give you clues to possible solutions?*

Developed countries are often huge energy users. They obtain the energy they need from a variety of sources including fossil fuels and hydroelectric developments. Yet fossil fuels are scarce and nonrewable and many countries lack the steep slopes and deep river basins necessary for efficient hydroelectric development. These forces create important incentives for the development of alternative energy sources, and they also give clues to possible solutions: a country should seek technologies that allow it to make the best use of its own resources (in Denmark's case, good sites for wind generation) and avoid technologies that force reliance on foreign resources and governments.

3. *What governments are responsible for the issue? Whose laws may apply?*

In this case, the Danish national (federal) government clearly has a lead role, although local and regional governments may become involved in practical decisions like siting wind generators and protecting residential areas from noise and accident. Denmark has a political obligation to the European Union to fulfil its commitment to install 1,500 MW of wind power by 2005.

4. *Who has a stake in the problem? Who should be involved in making decisions?*

Other than the governments listed in step 3, possible stakeholders would be the residential neighbors of wind farms, the manufacturers of wind generation equipment, environmental non-government organizations, particularly those concerned with the protection of wildlife, and public and private agencies whose responsibilities include public safety. Consideration of a particular site may help to identify additional local stakeholders, for instance industries or airfields close to areas where wind farms may be developed.

5. *In the view of your decision-making group, what are the attributes of a satisfactory solution? In other words, when will you be satisfied that the problem is "solved"?*

A satisfactory solution is one that gives Denmark a sustainable, renewable supply of energy without political conflict or vulnerability to fluctuating costs and supply. A number of solutions may fit these criteria.

6. *How will you evaluate (test, compare) potential solutions?*

Depending on how your decision makers view this problem, potential solutions could be evaluated on the basis of their total cost, or on their total generation potential, or on their total pollutant contribution, or (perhaps most important) on their ease of implementation locally, regionally, nationally, and internationally. This case does not require sophisticated mathematical analysis or computer simulation, except perhaps in the details of siting for maximum efficiency and minimum impact.

7. *What are all the feasible solutions to the problem?*

If we begin with the assumption that Denmark is seeking alternative, renewable sources of energy, we can consider a long list of factors that might allow successful development of alternative energy sources. If, on the other hand, we assume that Denmark is obliged (for instance, because of its European Union commitment) to pursue wind power, then we should examine only factors that relate to the development of wind power. These factors could include avoidance of international conflict, cost and polluting potential, and acceptability of new technologies to the public.

8. *Which solutions work "best" in terms of the attributes you identified in (5)?*

The "best" solution for Denmark—the factors that facilitate implementation of wind power—will vary depending on the social, political, and economic climate in the country, and will therefore change over time. Good ways for students to assess the effectiveness of different solutions in a case like this are role play or panel discussions, where different viewpoints can be brought together against a backdrop of factual information on cost and environmental performance.

9. *Which solution will be easiest to implement?*

Analysis of steps 1 to 8 should have revealed areas where Denmark is particularly vulnerable or sensitive in terms of wind power. Depending on prevailing public opinion, these areas may be noise, safety, pollution, impact on birds, ability to meet the country's promises to the European Union, or other factors. Ease of implementation—probably Denmark's overriding consideration in this case—will depend on the country's ability to satisfy these fundamental concerns among stakeholders. The government may be prepared to pay a little more for a technology that will be safe, reliable and widely accepted. Implementation must also consider potential impacts on Denmark's neighbors and tourist population. The solution that has the widest support may be the easiest to implement and thus the "best".

10. *What steps are needed for successful implementation? Who will pay? Who will monitor progress?*

This answer to this question will—like the effectiveness discussed in step 8—depend on public opinion, available resources, and dominant stakeholders. It is likely that a good implementation plan will require the approval of the European Union and perhaps individual neighboring countries in addition to Danish stakeholders. The steps themselves are possibly less important than obtaining agreement from those who will be responsible for implementation, especially any necessary spending, legislative changes, public education and monitoring of safety, noise and pollution.

6. *Where Can I Learn More About the Ecosystem, People, and Culture of Denmark?*

The following sources provide information on wind power in general and on the implementation of wind turbines in Denmark, other European countries, and elsewhere in the world.

J. C. Berkhuizen and A. F. L. Slob. 1989. The impact of environmental aspects on wind energy in The Netherlands. In: D. T. Swift-Hook (ed.), *Wind Energy and the Environment*. Peter Peregrinus Ltd., London, U.K.

S. Cave. 1992. Denmark's wind power lessons. *Our Planet* 4(5): 18-19.

D. Denniston. 1993. Second wind. *World Watch* 6(2) (March-April 1993): 33-35.

C. Flavin and N. Lenssen. 1991. Here comes the sun. *World Watch* 4(5) (September-October 1991): 10-18.

D. Lindley and D. T. Swift-Hook. 1989. The technical and economic status of wind energy. In: D. T. Swift-Hook (ed.), *Wind Energy and the Environment*. Peter Peregrinus Ltd., London, U.K.

N. O'Neill. 1989. Legal aspects of wind energy exploitation: an EEC perspective. In: D. T. Swift-Hook (ed.), *Wind Energy and the Environment*. Peter Peregrinus Ltd., London, U.K.

S. Stern. 1994. The way the wind blows. *International Management* 49(2): 33-35.

J. L. Tsipouridis. 1989. Wind energy development in Greece. In: D. T. Swift-Hook (ed.), *Wind Energy and the Environment*. Peter Peregrinus Ltd., London, U.K.

"How can we tackle the problems of traffic congestion and urban air pollution in Bangkok?"

1. What Is the Background?

Thailand is one of Asia's fastest-growing countries. Its capital, Bangkok, is located roughly in the center of the country, in an area that would be considered densely populated even without the city's huge population. The attraction of this area lies in its rich agricultural potential, although industrial development in Bangkok and several other smaller cities has increased rapidly in recent years.

The city of Bangkok is built on the delta of the Chao Phraya River, about 30 km upstream from the Gulf of Thailand. When the city was founded in the late 18th century, natural waterways flowing through the delta provided reliable transportation routes through the city. Over the centuries, natural channels have been augmented by constructed channels, creating a network of canals that continue to serve private and commercial traffic.

The city has grown quickly over the centuries. In 1850, the population of Bangkok was estimated at 400,000; today it is almost 6 million in the city, close to 9 million including suburbs.

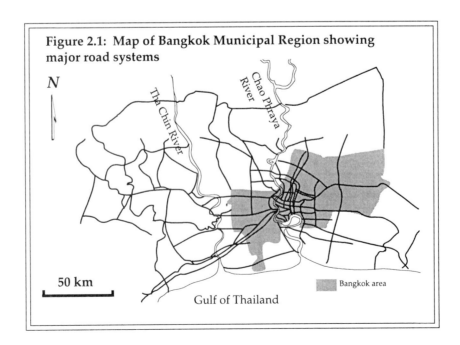

Figure 2.1: Map of Bangkok Municipal Region showing major road systems

Much of this growth has been unplanned and uncoordinated. As a result, people may build houses, and laneways to those houses, wherever they wish. Generally speaking, development is densest along major roads (so-called "ribbon development"), while narrow laneways lead back from the main road to parcels of land that have been divided and sub-divided many times. In some areas, lot sizes are now too small to support ordinary commercial uses or good residential development, and they become slum areas without adequate drainage, sanitation, or refuse collection. About one-sixth of the city area is slums.

The lack of clear land-use restrictions has also allowed the intermixture of different types of building, so that warehouses, factories, markets, racetracks, train stations, schools, and athletic facilities are built in among residential development. This combination of uses, and the network of small lanes leading to large roadways, considerably complicates traffic movement. Bangkok's rapid population growth, and even more rapid rate of motorization, has now created a situation where vehicular traffic and resulting air pollution are of serious concern. Today, Bangkok's traffic problems are considered to be among the worst in the world.

2. *What Problem Are We Trying to Solve?*

Bangkok's traffic congestion has a number of implications, some of which may not be obvious at first glance. The average speed of a vehicle in traffic is expected to dip below 8 km/hr by 2000, and it is not uncommon for a commuter to spend 6 to 8 hours a day in a car. In fact, each car in Bangkok is estimated to spend the equivalent of 44 days a year in traffic. Congestion is so heavy that parents of schoolchildren often leave home at 4 a.m., drive the children to school and sleep in the parking lot—all to beat the rush. FM 100, a local Bangkok radio station, plays traffic news 24 hours a day.

Traffic-induced delays have important economic implications, too. Business losses resulting from time lost in traffic jams are estimated to cost the city one-third of its potential gross city product, or about $4 million U.S. a day. If even 10% fewer vehicles were on the road during peak traffic times, city residents could save the equivalent of $400 million U.S. a year in lost productivity.

This level of traffic congestion cannot exist without significant environmental effects. The Thai National Environmental Board now believes air pollution to be the most pressing problem in the Bangkok metropolitan region. At least one in six Bangkok residents suffers from some kind of respiratory disease.

Motor vehicles are the major source of carbon monoxide in Bangkok, contributing more than 60% of the total. In dense traffic areas, carbon monoxide levels are well above the safe limit.

Table 2.1: Vehicle-related emissions in Bangkok: 1991

Carbon monoxide: > 1,000,000 T/yr
Nitrogen dioxide: 210,000 T/yr
Hydrocarbons: 150,000 T/yr
Sulfur dioxide: 100,000 T/yr
Suspended particulates: 80,000 T/yr
Lead: 600 T/yr

Vehicles also contribute high levels of hydrocarbons and nitrogen oxides. Ambient concentrations of lead in the air range from 0.1 to 1.0 µg/m^3—roughly three times the level considered safe in the United States. These levels are thought to cause 400 deaths a year, reduce children's IQs by an average of 4 points by the age of 7, and contribute to hypertension in 200,000 to 400,000 people a year. One estimate places the environmental costs of Bangkok's air and water pollution at $2 billion U.S. a year.

Even though the environmental and human health impacts of high vehicle density are clear, Bangkok's drivers seem resigned to the problem. In 1991, Thailand reduced duties on imported vehicles from 300% to 20 to 60%. At the same time, domestic vehicle production almost quadrupled from 1986 to 1992, when more than 300,000 vehicles were built. It is not surprising that the number of new car registrations in 1992 was 85% higher than in 1991. Traffic volumes typically increase 15 to 20% a year.

It is clear that Bangkok cannot continue to grow sustainably without addressing the problem of traffic congestion and its impacts. A comprehensive strategy is needed to reduce vehicular traffic and improve ambient air quality.

3. What Components of the Environment Are Affected, and How?

Development of the urban road network

Bangkok's urban mass now sprawls over a 50-km radius from the center. Because road networks have evolved without careful planning , some land areas are now isolated while others are experiencing dense development along the transportation corridors. Roads occupy only about 11% of the total land area, compared with 20 to 25% in other major cities like New York, London, or Paris.

The haphazard growth of the city has occurred in large part because anyone in Bangkok has the right to cut a new laneway or build a narrow path or street off a main road. As a result, many large roads have lanes entering them at 50- to 150-m intervals. Each lane may be only 4 or 5 m wide and 600 to 2,000 m long. There is no requirement to build sidewalks or provide drainage for these lanes, even though they may become feeder routes for the main road. In addition, parking in the city is unrestricted, with both sides of each major road lined with stationary vehicles, effectively reducing the number of traffic lanes by two.

The canal system that originally served the city, and a well-developed rail network, now carry little traffic. Instead, most commuters prefer to drive private cars. Car ownership is seen as an important status symbol in the Thai culture, and the comfort and convenience of private transportation are highly valued in comparison with mass public transit. Vehicle ownership in Thailand is now of the order of one vehicle for every 20 or so people. Although this number is small relative to the U.S. or Europe (rates of about 1.5 and 4 people per vehicle, respectively), it is far higher than in most other Asian countries. Indonesia, for example, has a ratio of about one vehicle for every 100 people; in India, there is only one vehicle for every 250 people.

Sources of air pollutants

Air-quality problems in Thailand arise from vehicular emissions, poor quality of fuel, inadequate vehicle maintenance, rapidly increasing industrial emissions, and poor industrial combustion technology. Although industrial emissions are important, the northwest-southeast orientation of the city of Bangkok and its location close to the coast encourage free air movement in outlying areas. In addition, a "green area" called Bangkrajao at the city's perimeter acts as a buffer zone to trap airborne pollutants.

Power generation and industrial sources contribute most of the city's sulfur dioxide emission and about half the airborne particulates; often, these facilities are located away from urban centers and therefore have limited impact on urban air quality. Nevertheless, two major power plants, together contributing about 11% of the country's total power production, are located within the Bangkok metropolitan area.

Clean-up priorities

Although other sources exist, it is clear that vehicular traffic is the single largest source of air pollutants in the Bangkok metropolitan area. It is also clear that current traffic density and transportation patterns have arisen as a direct result of government policies, or the lack thereof, relating to industrial and urban development.

To tackle the problem of air pollution in Bangkok, we will likely need to develop a plan with several components. (This discussion does not attempt to set clean-up priorities among the various air pollutants, just because vehicular traffic is so clearly the source of most in-city air problems in Bangkok.)

First, it will be necessary to clarify the roles and responsibilities of the various government agencies who are potentially stakeholders in the issue.

Second, we will have to develop alternatives to private vehicular transport. There are a number of proposals already available for this, so we may also have to develop ways of evaluating those proposals and deciding which is "best".

Third, it is probably desirable to develop a master plan for development of the Bangkok metropolitan

> *One of the greatest challenges in developing a mass transit scheme is diverting passengers from private vehicles to public transportation. Experience has shown that only about 5% are likely to make the change. Most mass transit passengers will in fact be diverted from other mass transit options such as buses.*

area, including a transportation component. This plan should accommodate not only present concerns but should also look into the future to consider what Bangkok might look like 10 or 20 years from now. In addition to a vision of the future, the plan should contain specific guidance on issues like land use, areas to be developed versus retained as green space, desired population density, and transportation corridors.

Finally, it may be important to reconsider economic policies relating to the production, import, and sale of vehicles, fuels, and emission control devices.

Government agencies and responsibilities

There is no central agency charged with overseeing planning activities in Bangkok or in Thailand overall. Some of the agencies with an interest in the problem are:

Bangkok Metropolitan Authority (BMA) (municipal)
Expressway and Rapid Transit Authority (municipal)
State Railway of Thailand (national)
Department of Highways (national)
Ministry of Public Health (national)
National Environmental Board (national)
Department of Industrial Works (national)

Because responsibility for planning is not always clear, attempts to fix the problem have arisen on many fronts and have sometimes come into conflict. At least 5 separate mass transit "megaprojects" have been proposed (see Box 2.1), even though building all 5 would not be possible simply because their physical locations conflict. In other cases, there would be needless duplication of facilities and even of services.

The list of "megaprojects" shows a strong emphasis on two elements: an elevated rapid transit system and a network of high-capacity roadways. Neither of these approaches, on its own, is a sufficient solution to the problem. The reasons for this illustrate the need for centralized and well-coordinated planning of major municipal works.

The need for local infrastructure

The major expressway projects underway or proposed for Bangkok will vastly increase the number of cars that can travel comfortably in key transportation corridors. But how will those vehicles enter and leave the expressways?

The Don Muang Airport highway currently has 10 lanes (5 in each direction), with overpasses to serve on- and off-ramps. The proposed work will expand the system to 26 lanes by adding a 6-lane elevated tollway, and the Hopewell Project's 6-lane highway and 4-lane local road system.

At present, traffic tie-ups on the Don Muang Airport highway occur not on the highway itself but at junctions, where slower moving traffic is entering or leaving the expressway. To resolve the overall congestion problem, planners must not only build more expressway capacity but must also improve the junction system and the number and size of local roads feeding the expressways. To be effective over the longer term, an adequate transportation plan must resolve current problems while providing for future expansion in the region over at least the next 10, preferably 20, years.

The challenges of urban in-filling

Bangkok is an old and densely populated city. Most of the proposals to build mass transit facilities incorporate an elevated system of tracks running through the downtown core. If even two of the proposed systems are built, there will likely be a need to accommodate crossovers of the two track systems at some height above the urban street. Some analyses show that the three main urban transit plans conflict or overlap directly at 33 separate locations within the city.

If an expressway is also present, there is the potential for a stack of at least three transportation structures, the uppermost of which would be several tens of meters above the ground. The aesthetic, noise, and dust impacts of such an arrangement would be considerable—not to mention the engineering challenges of building stacked stations to serve the two or more rapid transit systems. In some key locations, all three systems would potentially be present, requiring stations as high as 30 meters.

The need to reduce vehicular traffic

Building more roads may make the commuter experience more pleasant, but it will not ultimately solve the problem of vehicular emissions and their environmental and human health impacts.

One obvious solution is to promote public transportation as a viable alternative to private vehicle use. Major challenges exist here, especially in the attitudes of drivers who prize their cars and are resigned to the congestion and smog. It is expected that the opening of mass transit routes will also release pent-up demand from people who want to travel but do not own a car or prefer not to drive—in other words, new customers, not converts from private vehicles. Experience from other cities suggests that only about 5% of car owners will opt for public transit; most other transit riders will have formerly used other mass transit options such as buses.

4. How Can I Analyze This Information?

Clarify the goals

Often, environmental decision-making fails because the goals of a policy are not clear. Before beginning an analysis of solutions, it is important to have planning goals clear and agreed to by all participants. In the case of Bangkok's traffic congestion and consequent air pollution, an appropriate goal might be to reduce vehicular traffic overall, rather than to expand capacity. If achieved, the former will reduce congestion while also reducing air pollution; the latter will reduce traffic jams but may actually increase air pollution if other emission controls are not imposed.

If addition of capacity is determined to be advisable (for whatever reason), it will be important to be clear about where capacity should be added, and in what way. This becomes easier if a system-wide analysis is undertaken (see below). Major projects like the Don Muang Airport Tollway will more likely be successful if supporting infrastructure is considered as part of the overall analysis.

Assign the responsibilities

The lack of clear roles and responsibilities for Thai national and Bangkok municipal agencies has been a key source of confusion in the management of transportation problems and their solutions. Most of Bangkok's megaprojects have been plagued with delays, contract disputes, rising land costs, and financing. Political rivalries between Thai politicians further complicate the decision process.

It will not be easy to resolve political difficulties or establish clear jurisdiction (responsibility) over aspects of the planning problem. Yet these forces may currently be the most important obstacles to an integrated transportation planning scheme for Bangkok.

Establish the legislative framework

Once jurisdiction has been established—which agency has which responsibilities—it will be possible to enforce, review, and perhaps restructure the laws and policies that control land-use planning in general, transportation planning in particular, vehicular emissions and their control, and the importation and production of new vehicles.

Various controls are possible, ranging from control over automotive emissions, to fuels allowable for sale, and even to the technologies permissible for new vehicles. Box 2.2 illustrates the approach taken in Hong Kong, which has experienced similar rapid growth and traffic congestion in recent decades.

Box 2.2: Traffic solutions for Hong Kong

Hong Kong is one of the world's busiest and most densely populated cities. Private car usage in Hong Kong more than doubled in the 20-year period from 1971 to 1991, while truck traffic increased almost sixfold.

A 1989 white paper (government policy proposal) presented a coordinated strategy for improving air quality and relieving traffic problems in Hong Kong. Its key recommendations are to:

1. Divide Hong Kong into 10 air-quality zones to facilitate air-quality monitoring and control.

2. Require newly manufactured engines to comply with strict vehicle emission standards consistent with prevailing European standards. Aim to strengthen these standards over time.

3. Enforce strict vehicle emissions standards on "in-service" vehicles. Cars and trucks reported to be emitting "black smoke" would be required to report for reinspection at certified test centers and would not be permitted back in service until necessary repairs have been completed.

4. Appoint official "smoke spotters" to monitor emissions of in-service vehicles operating in traffic.

5. Introduce unleaded gasoline as soon as possible, preferably at a lower price than leaded fuel. Unleaded fuel is also a requirement for the catalytic converters that must be installed to meet other emissions limits.

6. Extend the mass transit system, especially light rail transit and the Mass Transit Railway.

7. Encourage bus companies operating in congested areas to replace diesel engines with electric ones.

8. Encourage taxis and public light buses to convert to liquid propane gas fuel (although safety is an issue of concern in this option, particularly in Hong Kong's several major tunnels).

Develop a strategic plan

The next step in a solution is to develop a long-term plan that will accommodate not only existing problems but also future development.

Such a plan requires the choice of several important variables. Among these are a suitable planning horizon, typically at least 10 years and often as long as 50 years.

An appropriate discount rate must also be selected for any economic analysis. The discount rate is the rate at which money loses buying power with time. It is usually a combination of influences including projected inflation, interest rates, and similar factors.

Physical boundaries for the analysis must also be chosen: Should it be limited to the Bangkok Metropolitan Region? Or include some outlying areas or particular towns?—and so on.

Finally, the plan must have the support of the community and of the politicians who in the long run will be responsible for overseeing its implementation and possibly its financing. Thus there is a need for a comprehensive consultation framework within which public and political opinion can be sought and consensus developed.

Develop alternative solutions

Within the context of a strategic plan, there is still flexibility to develop short-term strategies to deal with specific aspects of the problem. In the case of Bangkok, it certainly makes sense to develop some alternatives to private vehicle use. At present, the focus seems to be on light rail transit within the city, coupled with commuter rail systems linking Bangkok to neighboring towns. There may be other approaches to this, including expansion of the bus system, but these may be more difficult to implement if they depend again on a less-than-adequate road network.

The key here is to develop more than one solution that will satisfy the goals of the long-term plan. More than one solution allows flexibility and leaves the door open to varied input from stakeholders. You may also need different solutions for different periods of time in your strategic plan—one for the short term, one to be implemented 5 years from now, and one long-term "ideal" solution, for example.

Developing several feasible solutions—at least on paper!—also allows you to accommodate changing conditions as the future unfolds. A strategic plan gives you general guidance over a long period of time, but it must be reviewed and updated frequently as external conditions change. Having several practical solutions available allows the planner to respond quickly while retaining consistency with the overall plan.

5. *How Can I Use My Findings to Reach a Solution?*

The complexity of Bangkok's physical structure, and the interlinkage of environmental with land-use planning issues, makes this a daunting case for any environmental analyst. The problem-solving framework described in the Introduction can help you clarify your thinking and focus your analysis on the most crucial elements of the problem.

1. What is the problem?

In Section 2, we identified a need for a comprehensive strategy to reduce Bangkok's vehicular traffic and improve ambient air quality.

2. In what ways do human activities have impact on the natural environment to cause "a problem"? How do these mechanisms give you clues to possible solutions?

Although a number of human activities may affect air quality in Bangkok, available data show that vehicular traffic is the city's main source of air pollutants. Air pollution from vehicles occurs because of poor vehicle maintenance, long idling times due to traffic congestion, variable fuel quality, and other factors. This situation clearly directs us to solutions that (a) reduce the number of vehicles on the road; (b) reduce the time each vehicle spends in traffic each day (for instance by improving road networks); and (c) improve the quality of emissions from each vehicle (for instance by imposing restrictions on vehicle certification and maintenance requirements and by improving the quality of fuel). It is clear from the complexity of the problem that there is no "quick fix" available. A good solution will have a number of components which will likely involve land-use planning issues, restrictions on vehicle and fuel importation and use, and provision of good public transportation. The current resistance of many Thais to public transportation could suggest the need for a strong public education program.

3. What governments are responsible for the issue? Whose laws may apply?

Jurisdictional confusion is central to this problem. It is not at all clear who should be doing what or even which government agency has a lead role. This confusion is certainly part of the problem, and resolving that confusion should be a target of any good solution.

4. Who has a stake in the problem? Who should be involved in making decisions?

Aside from the commuters and the non-commuting residents (such as children) who experience impaired air quality, important stakeholders include the various government agencies with an actual or potential interest in the problem, the proponents, both public and private, of the various transportation "megaprojects", and the manufacturers of public transportation equipment, private vehicles, fuels, and pollution control equipment for vehicles.

5. In the view of your decision-making group, what are the attributes of a satisfactory solution? In other words, when will you be satisfied that the problem is "solved"?

There are at least two ways to answer this question. One is the very literal interpretation that air pollution problems in Bangkok will be "solved" when the levels of key pollutants drop below those recommended by agencies such as the World Health Organization. A second interpretation could include broader considerations such as land-use planning, for instance that the problem will be "solved" when Bangkok not only meets air-pollution standards but also has a sound long-term plan for urban development, to ensure long-term reductions in traffic congestion and thus long-term improvements in air pollution. The choice (and there are other possible "solutions" here) will depend on the values and priorities of the decision-making group.

6. How will you evaluate (test, compare) potential solutions?

Many analytical methods could be useful in this case, depending on the priorities established in step 5. Computer simulation could assist in predicting the concentrations of air pollutants and the density of traffic, both under present circumstances and under future management scenarios. Cost-benefit analysis could be helpful in deciding which combination of the megaprojects makes best economic sense. Analysis of epidemiological evidence from other urban centers could help us understand the impacts on human health of various possible management actions. As with other studies, it may make sense to use a combination of methods to decide which management options are "best".

7. *What are all the feasible solutions to the problem?*

We know that this is a complex problem, but it is not a unique one. We could develop a list of management options by looking at case histories from other centers like Hong Kong, New Delhi, Jakarta, Cairo, New York and London. Which strategies have worked? Which haven't? What elements do the successful strategies have in common? But don't stop there: Bangkok is not Hong Kong. Use imagination and first principles to develop other possible solutions. Then exclude those which may be difficult to implement to create a list of feasible options.

8. *Which solutions work "best" in terms of the attributes you identified in (5)?*

Now see how each of your feasible management actions shapes up against the criteria you selected in step 5. Which pollutes least? Or, which encompasses long-term land-use planning? Which costs least? Which is least complex in terms of administrative systems? And so on. Even a superficial analysis will quickly reveal that certain options just won't work as well as others. These can then be excluded from further analysis, leaving you with a short list of workable solutions. These may be more or less equivalent in terms of cost and pollution, for instance, but may differ in their ease of implementation because of factors like administrative complexity. The decision-making group must ultimately decide which of your short list is "best" based on these factors.

9. *Which solution will be easiest to implement?*

The lack of clear administrative responsibilities (jurisdiction) makes this case particularly difficult to resolve. Clearly, ease of implementation will be a central concern. This ease could be achieved by simplifying the roles of the different agencies and perhaps entrenching the new roles under the law. But legislative change takes time. It may be that there is one agency that has the longest-standing interest in the issue of air pollution, or land-use planning, and which could therefore become the lead agency for an implementation scheme. Or it may be that some agencies, although interested in the problem, have little money or personnel to contribute to it. If so, it would not make sense to assign heavy responsibility to such an agency. Public-private partnerships, such as are underway in some of the megaprojects, can help smooth implementation by easing the problem of financing; they may also, however, have their own implementation problems if several agencies disagree about the scope of their activities.

10. *What steps are needed for successful implementation? Who will pay? Who will monitor progress?*

With an issue as complex as this, it makes sense to set out a long-term plan with several interim "check points" along the way. A detailed implementation plan should therefore include provisions for program re-evaluation (and, if necessary, reconfiguration) at, say, 3-year intervals. Such a system allows decision makers to adapt to changing economic and societal conditions, and also to make best use of new technologies as they become available on the market.

6. *Where Can I Learn More About the Ecosystem, People, and Culture of Bangkok?*

The following sources contain a variety of information about population, transportation and urban air pollution issues in Thailand and other Asian countries.

C. S. Cheung and M. J. Pomfret. 1993. The management of motor vehicle air pollution in Hong Kong. In: B. Nath, L. Candela, L. Hens, and J. P. Robinson (eds.), *Environmental Pollution—ICEP 2.* European Centre for Pollution Research, London, U.K.

M. Lowe. 1991. Calming motorized traffic. *Alternatives* 18(1): 16-17.

P. Midgely. 1994. Urban Transport in Asia: An Operational Agenda for the 1990s. World Bank Technical Paper Number 224. World Bank, Washington, D.C.

Organisation for Economic Co-operation and Development. 1992. Market and Government Failures in Environmental Management: The Case of Transport. OECD, Paris, France.

D. Phantumvanit and W. Liengcharernsit. 1989. Coming to terms with Bangkok's environmental problems. *Environment and Urbanization* 1(1): 31-39.

R. Strickland. 1993. Bangkok's urban transport crisis. *The Urban Age* 2(1): 1-5.

Washington

"How can we reduce discharges of toxic chemicals from small industries to local receiving waters?"

1. What Is the Background?

A growing community with growing impacts

The southern portion of Puget Sound, Washington, is one of the fastest growing urban areas in the state of Washington. New residents are drawn to the area by its abundance of lakes, streams, and wetlands and its valuable groundwater supplies for drinking water.

The area has experienced many changes with this rapid development. Among them is the increase in the number of small industries supporting the community, especially construction, landscaping, and automotive repair services. As the area has grown, new city ordinances and restrictions have also been adopted, in an effort to control the impacts of development and protect valued resources. Some of these new local laws are local initiatives; others are the result of state or federal requirements. Regardless of their source, these regulatory changes have the potential to increase the costs of doing business. Furthermore, few small industries have the capability to track changing regulatory requirements, so getting industries to comply with the new rules may be difficult for the agencies enforcing them.

Add to this mixture local initiatives for whole-watershed management, and a very confusing picture begins to emerge. This is the stuation that confronted the neighboring cities of Olympia, Thurston, and Lacey, Washington, in the early 1990s. Aided by grants from state and federal agencies, the three

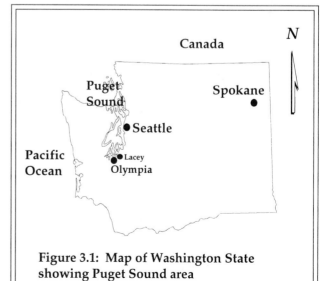

Figure 3.1: Map of Washington State showing Puget Sound area

cities began a new cooperative initiative that they termed Operation: Water Works to raise awareness of the need for water stewardship within the business sector and in the community as a whole.

In the summer of 1991, the three cities began to develop a project focus, a detailed work plan, and the materials they would need for the project. They also established a comprehensive consultation program, a regional steering committee, and a Business Advisory Committee to direct the work as it proceeded.

[?] 2. What Problem Are We Trying to Solve?

Within any urban area, there are potentially innumerable environmental "problems" to deal with. The first challenge is therefore to decide which problems are most pressing—which should be resolved first. There are a number of ways to arrive at this decision. Case Study 6 gives some examples for the Ganges River ecosystem in India. Case Study 4 shows how a multiobjective planning process could be structured in New Zealand.

For the cities of Olympia, Thurston, and Lacey, a key objective was the creation of a sense of partnership and coordination between the business community and government. They also wanted to establish a system that would be preventive rather than reactive, and encourage businesses to become environmental stewards rather than adversaries of government.

The cities chose to develop a central goal for their joint initiative through discussions with the project Steering Committee and the Business Advisory Committee (see membership list in the box on the next page). Within their overall goal—to achieve water and environmental protection and stewardship through proper management of hazardous materials and business operations and maintenance—they set specific objectives. These objectives included:

1. A cooperative and constructive approach

2. An ethic of water/environmental stewardship

3. Consideration of surface water quality and hazardous material concerns

> *Operation: Water Works' primary goal was to achieve water and environmental protection and stewardship through proper management of hazardous materials and business operations and maintenance.*

But what "problem" is really being solved here? The Steering and Business Advisory Committees recommended an emphasis on protecting the area's treasured water resources and on hazardous wastes as the greatest threat to those resources. Another community, in another location, might arrive at a different focus for work. What may be most important is that the decision-makers—who, as elected representatives, reflect the wishes and attitudes of the community—agree on the chosen priorities.

Using their project goal as a guide, the committees decided to screen local activities to decide which had the highest potential to pollute local lakes, streams and wetlands. They used three main criteria in this screening:

1. The *quantity of hazardous materials or wastes* the activity handled or produced

2. The potential water quality and environmental problems that would be created with *the disposal of the activity's wastes*

3. The potential water-quality and environmental problems associated with the activity's *routine operation and maintenance*

Preliminary surveys of area businesses revealed that the industries which had most potential to use or dispose of hazardous materials, and thus to affect the quality of surface waters or other components of the environment, were automotive-related businesses, equipment manufacturing and repair businesses, landscaping businesses, construction companies, and building maintenance and cleaning related businesses. (In another community, the mix of businesses considered to be "most of concern" would likely differ from this list.)

We can therefore state the problem that we are trying to solve as "How can we achieve water and environmental protection and stewardship through proper management of hazardous materials and business operations and maintenance in automotive-related businesses, equipment manufacturing and repair, landscaping, construction, and building maintenance and cleaning?"

 # 3. What Components of the Environment Are Affected, and How?

The target industries

Identifying a group of target activities allowed the project team to focus their limited time and financial resources on a few key activities. In Lacey, the project team chose to identify these activities by matching their Standard Industrial Classification (SIC) code listings from the Department of Revenue with local business addresses. They also used local business licenses to help find businesses in the target categories. In Olympia, project team members conducted a "windshield survey"—driving around the survey areas and looking for businesses that appeared to fall into the target categories.

Throughout the study area, businesses were identified through listings in the local telephone directory and through mail, phone, and on-site surveys.

Other approaches could have employed local municipalities' knowledge of their industrial community and potential polluters, but the results would likely have been less specific than were obtained through surveys.

Box 3.2: Operation: Water Works priority business categories

Business type	Waste-generating activities
Automotive businesses General auto service Auto repair Vehicle storage Wrecking yards	Vehicle washing; fueling; storage; sanding, blasting, grinding; use of paints, solvents, and oils; working close to lakes, streams, and wetlands.
Landscaping Nurseries and turf farms Lawn care Landscape architects Landscape contractors	Selecting and installing plants and other landscape materials; maintaining landscapes; selling plants, fertilizers, and pesticides; working near lakes, streams, and wetlands.
Construction New site construction Maintenance and remodelling Painting Cement suppliers and installers	Clearing and grading land; pouring concrete; painting, sanding, and blasting; using paints, solvents, and adhesives; using batteries; working near lakes, streams, and wetlands.
Building maintenance and cleaning Office cleaning Building management Carpet, upholstery cleaning Window cleaning	Parking and loading; handling chemical cleaners, preservatives, sealers, solvents, and oils; storing wastes and materials; producing wastes (e.g., carpet cleaning sludges, oil, contaminated chemicals, debris); washing equipment, vehicles, buildings, roofs, pavement; sealing driveways; working near lakes, streams, and wetlands.
Equipment manufacturing and repair Heavy equipment repair (e.g., construction equipment) Small equipment repair (e.g., lawn mowers, appliances) Boat manufacturing and repair	Washing mechanical equipment; storage of outdoor equipment and supplies; sanding, blasting, and grinding; use of paints, solvents, oils, gasoline, and detergents; producing and handling wastes such as paints, solvents, used oil, used gasoline; storing materials and wastes; working near lakes, streams, and wetlands.

The surveys revealed a total of 2,719 Thurston County businesses in the target industry groups. Figure 3.2 shows the relative proportions of these businesses.

Businesses were then asked (in telephone, mail, and on-site surveys) what practices they currently employ to handle hazardous waste and preserve water resources. For most of these businesses, stormwater control (management of the runoff from roofs and parking lots) was the major water stewardship activity. Although very few businesses were found to be using practices that were immediately threatening to the environment or human health, at least a quarter of the businesses surveyed were using poor management practices. For example, 84% did not have measures in place to collect and treat stormwater runoff and 68% were using their floor drains improperly. Most businesses needed to make changes of one kind or another, whether structural (e.g., install stormwater collection and retention facilities) or operational (e.g., sweeping surfaces clean rather than hosing them off).

Why were poor management practices continuing, even when businesses were aware of their potential impacts on water quality and the environment? The study team found several major reasons:

1.   ***Businesses lacked the incentive to change***, because local and state agencies didn't monitor the use of good practices, because the business operator was unaware of the impact of current practices, or simply because making the change would be costly and difficult, requiring changes to the physical structure of the facility.

2.   ***"Grandfathering" shelters offenders.*** When new ordinances are brought in, old practices already in place are often allowed to continue. This so-called "grandfathering" makes it difficult to find and educate businesses that may be significant contributors to environmental degradation.

3.   ***Convenience and lower costs encourage proper management.*** It is often the case that careful storage and reuse of materials reduce costs and thus make good business sense. But where an action is costly or inconvenient, a business is more likely to use poor management practices.

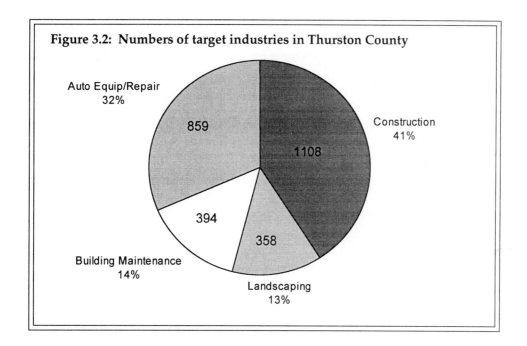

Figure 3.2: Numbers of target industries in Thurston County

Auto Equip/Repair 32%
859
Construction 41%
1108
394
358
Building Maintenance 14%
Landscaping 13%

4.   ***Intermittent wastes are perceived as expensive to recycle.*** Where wastes are generated infrequently—perhaps only once a year—it can be hard to convince a business that it's cost-effective to dispose of them properly.

5.   ***Businesses need better education on management options and regulatory requirements.*** The most frequently asked questions in the target industries were "what regulations apply to me?" and "what is the appropriate way to dispose of the wastes I generate?" Without good information, businesses cannot be expected to adopt appropriate management practices or even to understand the environmental impact of their current activities.

4. How Can I Analyze This Information?

What practical steps are needed?

Operation: Water Works set an overall goal for protection of Thurston County's valued water resources, identified key activities most likely to have impact on those resources, and determined a number of reasons why businesses were not currently using environmentally protective practices. In other words, they defined the problem and its sources. What practical steps should now be taken to encourage businesses to change their ways and thus reduce impacts on the environment?

To answer this question we must examine the results of business surveys and the real or perceived obstacles to change.

One of the first things that is apparent from the surveys is that businesses simply need better information, so an appropriate project activity would be some sort of educational program.

It was also clear that each individual business is in some way a special case, whether because of its raw materials, its customers, its physical situation, the layout of its facility, or some other factor. So there is probably a need to "customize" remedial actions on a case-by-case basis. Depending on their circumstances, and perhaps their history of pollution, industries may be suspicious of a government initiative to change the way they work. Others may be confused about the goals of the project or even the steps a business has to take to become a participant. There may therefore also be a need for on-site, face-to-face contact to provide site-specific guidance and reassure participants.

Third, businesses were understandably concerned about profitability and their profile in the community. Cooperation might be encouraged by some sort of reward or public recognition for participants.

What about monitoring and enforcement?

The original plans for Operation: Water Works called for the agencies to monitor water quality, use monitoring results to identify pollutant sources, identify polluting industries, and enforce clean-up through formal Notices of Violation.

As the project evolved, however, it became apparent that this approach had two main drawbacks. First, it would be costly to maintain an appropriate level of water-quality monitoring—and project funds were very limited. And second, it would be difficult to maintain a cooperative and

---

Box 3.3: Cost sharing arrangements for Operation: Water Works

The costs of this project were borne cooperatively by several agencies. The Cities of Olympia and Lacey each received grants from the state Centennial Clean Water Fund to provide local businesses with information and education on waste management and water quality impacts.

Olympia's grant focused on developing best management practices for key industries. Lacey's grant was targeted at identifying and correcting the main sources of pollution to storm drainage systems.

Project activities were carefully linked to routine city initiatives such as business inventories, health department outreach, and facility design.

voluntary approach if businesses were constantly in fear of strict enforcement action.

The discussion then led to a central question: should the project spend its limited funds on monitoring suspected problem areas or on gathering more general information that would be helpful to local businesses?

After much consideration, the Steering Committee elected not to use monitoring as part of the project. Their decision was based on the high cost and uncertainty of an effective monitoring program, the potential of monitoring to create an adversarial relationship with business, and the need to create effective business education and technical assistance programs rather than "punishing" offenders.

The Steering Committee also developed a set of guidelines for site visits. Staff visiting a business would explain the goals of the project and the need for environmental protection. The business operator would be made aware that Operation: Water Works is not an enforcement program but rather a support and education initiative. Where staff observed an obvious violation of local ordinances, the business would be informed of this and asked to work out an agreement with project staff to solve the problem. Project staff would then work with the business and provide liaison with regulatory agencies to arrive at a solution without resorting to enforcement action.

> *Operation: Water Works intentionally did not set targets for reductions in key pollutants. Instead, they chose to target types of activities that were likely to contribute to environmental degradation.*
>
> *This approach has the advantage of being simple to understand and implement. It does however have the disadvantage of not guaranteeing a particular outcome in terms of pollutant reductions or environmental quality.*

Does this mean no evaluation is necessary?

Not at all! But project staff chose to focus their evaluation on the success of the various public involvement and education methods they used, assessing outcomes against the goals and objectives of the project.

In conducting this evaluation, project staff examined Operation: Water Works activities at seven levels of evidence, as follows:

1. *Inputs*—the staff time, volunteer time, money and materials that contribute directly to a sponsor's ability to offer educational or public involvement...

2. *Activities*—where activities are targeted at particular...

3. *Participants*—Participation by key players leads to personal or corporate...

4. *Reactions*—and these reactions contribute to the participants' aquisition of new...

5. *Knowledge, attitudes, skills and aspirations*—Changes in knowledge, attitudes, skills and aspirations then lead to observable...

6. *Practice changes*—and these practice changes, changes in behavior, ultimately produce the desired...

7. *End result.*

Each of these seven "evidence levels" can be measured and assessed. Although Operation: Water Works was not set up this way, the seven levels can also be used to set specific performance targets. For example, *input* goals could be specified before the project begins: how many people are needed, how much money, where the money will come from, and so on. Desired *reactions* can also be specified: for example, that businesses will be pleased and surprised by the impact they can have on environmental quality through modest operational or structural changes.

Although the project was not in fact structured this way, its outcome can be evaluated on the seven levels. The following summarizes these results.

Inputs

The project used 12 staff people from the participating agencies and another 12 contractors as trainers, consultants, writers, editors, and graphic designers. A total of $235,725 was budgeted for the project, of which the Clean Water Fund grant accounted for $176,794. The City of Olympia, the Thurston County Storm and Surface Water Program, and the Thurston Country Environmental Health Program each contributed $18,000. A portion of the City of Lacey's staffing expenses (including a cash amount of $13,000) was funded by a separate Clean Water Fund grant.

Although these levels were adequate for the task (and in fact the project ran slightly under budget), it was felt that some efficiencies could have been achieved by hiring a single writer for the project (rather than several separate consultants) and by seeking less expensive ways of producing printed materials. Brochures could take the place of some lengthier publications. Remaining project funds will be used to print such a brochure and provide additional one-on-one support for businesses.

Activities

Operation: Water Works conducted four kinds of educational and outreach activity: developing educational materials, publicizing and promoting the program, conducting events such as workshops, and conducting Pollution Prevention Plan consultations with individual businesses.

Assessment of these activities revealed a number of issues. It was difficult to be sure that survey data were correct or even uniform, and response rates on surveys were often low. A better survey design might have enabled staff to collect more specific responses. On-site surveys were more effective for collecting information because staff could observe business practices. On the other hand, to conduct an on-site survey, it was necessary to obtain the operator's permission to enter the site and open their records, and sometimes this permission was difficult to get.

Printed materials were time-consuming to prepare and costly to produce. Certain elements did, however, work very well, particularly postage-paid reply cards inserted in brochures for workshop registration, and community recognition advertisements. Feedback from businesses suggested that outreach activities, such as presentations to community organizations like Rotary Clubs, were very important in dispelling initial distrust and skepticism and giving business owners the chance to talk to staff on "neutral ground."

The workshops were considered to be very successful, giving business operators the chance to exchange ideas and share information in a nonthreatening environment. Reminder calls to preregistrants helped to boost attendance at the second round of workshops. Timing of the workshop (time of day and time of year) was very important in securing good attendance.

> **Box 3.4: Operation: Water Works project activities**
>
> 1. *Five news releases, generating 25 mentions or articles in 12 local or national publications.*
>
> 2. *Twelve presentations to local civic, business, environmental, and governmental organizations.*
>
> 3. *Representation at 10 conferences.*
>
> 4. *Five quarter-page ads in the* **South Sound Business Examiner.**
>
> 5. *3,100 brochures mailed to Thurston County businesses in 1992; 3,200 in 1993.*
>
> 6. *250 copies of the project Handbooks and Self-Assessment/Pollution Prevention Plan forms distributed.*
>
> 7. *250 window decals printed.*
>
> 8. *Two rounds of workshops, with one workshop for each business category in each of 1992 and 1993. Participation was 124 pre-registered participants in 1992 and 108 in 1993. Note that in each year a significant number of registered businesses did not in fact attend the workshop (70 did not attend in 1992; 24 in 1993). Information packages were mailed to those interested in information but unable to attend.*
>
> 9. *45 Pollution Prevention Plan consultations resulting in 38 Pollution Prevention Plans being completed.*

Pollution-prevention planning faced major challenges in this project. Businesses were often very busy during office hours and were unavailable for one-to-one consultation. Although self-assessment forms were prepared for businesses to use, in fact project staff spent much of their time working through the forms with interested businesses because of questions that arose during the evaluation process. The pollution-prevention-planning process was intended to be flexible to allow businesses to be creative across the full range of their activities; however, its lack of structure was in fact confusing for many operators. Some suggested that they would prefer a form that listed candidate best management practices and offered specific advice about what changes might be feasible. Actually getting businesses to complete the plan (rather than just to begin it) was in itself a challenge because of conflicting time demands. As a result, many businesses who completed only part of the planning process did not receive community recognition for their efforts, however good.

> *Outreach activities, such as presentations to community organizations like Rotary Clubs, were very important in dispelling initial distrust and skepticism and giving business owners the chance to talk to staff on neutral ground.*

Participation

In the end, 39 businesses of a possible 3,300 in Thurston County completed pollution prevention plans and 141 business operators attended pollution prevention planning workshops. Although this seems like a very small proportion of the total, it compares well with experience from other programs elsewhere. For instance, the nearby city of Bellevue, which has a population of 95,000 people, had 55 business participants and 450 workshop participants in a similar exercise after 2½ years of education and outreach.

Project staff and local businesses agree that participation would have been higher had more effective enforcement been in place, for instance if the project had been coordinated with the adoption of a new city ordinance. Telephone follow-up was very important for encouraging participation, but additional incentives and approaches might further improve attendance. In the landscaping workshops, trainers were able to offer Pesticide Applicator Recertification credits through the state Department of Agriculture.

Reactions

Reactions to the workshops were very positive, probably because the workshop format allows skeptical business operators to learn about the project without making a formal commitment of time or money. On their own, the workshops did not appear to be effective in motivating businesses to begin pollution prevention planning. Nevertheless, a great many local businesses now have project materials and the combination of telephone follow-up, workshops, and printed materials have created a more receptive climate for discussion of pollution-prevention and environmental stewardship in the business community.

> *Environmental monitoring was not within the scope of Operation: Water Works, so staff were unable to measure actual improvements in environmental quality as a result of project activities. They believe, however, that the combination of education, outreach and technical assistance will change attitudes and practices and thus reduce environmental impacts over the long term.*

Change of knowledge, attitudes, skills, and/or aspirations

Assessments of changes in basic knowlege, attitudes, skills, and aspirations as a result of project activities clearly shows that businesses who participated in workshops gained knowledge. Those who completed self-assessments learned even more about themselves.

The project's ability to evaluate the learning experience was limited, however. Better pre- and post-workshop survey questions could shed light on participants' knowledge on arrival versus after a workshop. Aspiration and attitudes can be tested before and after the learning experience in the same way.

The project gave rewards (public recognition) for a business's willingness to think about its environmental practices but did not require actual implementation for that reward to occur. Nor did the project follow up consistently with businesses that had completed pollution prevention plans, or even whether they had posted community recognition decals (thus encouraging public response to their behavior). Businesses often commented that they had expected such follow-up visits and felt that a 6-month visit would be most helpful in encouraging completion and implementation of the pollution prevention plan.

Practice change

The goal of Operation: Water Works was education and technical assistance. Changes in business practice were desirable, but the project's lack of a strong enforcement component (or strong incentives) made it difficult to guarantee that those changes would occur. Project results clearly showed that businesses were confused about their regulatory requirements, about their pollution prevention options, and about the potential of their practices to impact the environment. Many felt that it was better to take the risk of violating local ordinances rather than take the time and money necessary to figure out what they had to do to be in compliance.

In retrospect, project staff believe that a "stick" (enforcement) component should have been added to the "carrot" of technical support. The incentives for change were not clear for most businesses, so participation was cavalier and actual practice changes were more limited than might have been the case with strict enforcement of ordinances.

End result

In assessing the end result of the project, it is necessary to evelute the actual changes that have occurred in the environment as a result of project activities. Appropriate measures of these changes might be decreases in nonpoint source pollution (i.e., agricultural and urban runoff) entering local streams, groundwater and Puget Sound, and decreases in sedimentation in local streams. But because environmental monitoring was intentionally excluded from

> *In retrospect, project staff believe that a "stick" (enforcement) component should have been added to the "carrot" of technical support.*

the project, project staff were unable to identify any measured effects that were directly related to practice changes. As mentioned above, monitoring is a costly and time-consuming process and thus a major project commitment, which the Steering Committee chose not to endorse. Business participants agreed with this approach, believing that monitoring was likely to be a waste of precious resources and not necessary from their perspective.

Project staff believe that the combination of education and technical support offered by Operation: Water Works has led or will lead to changes in attitudes and changed practices and that these changes, however small they may be, will reduce impacts on water resources and environmental quality.

5. How Can I Use My Findings to Reach a Solution?

Use the decision-making framework described in the Introduction to organize your thinking on this problem, as follows:

1. What is the problem?

In Section 2, we identified the problem as "How can we achieve water and environmental protection and stewardship through proper management of hazardous materials and business operations and maintenance in automotive-related businesses, equipment manufacturing and repair, landscaping, construction, and building maintenance and cleaning?"

2. In what ways do human activities have impact on the natural environment to cause "a problem"? How do these mechanisms give you clues to possible solutions?

This case describes a number of human activities (see especially Box 3.2) that impact directly on the aquatic environment. It is very difficult to recover spilled materials like solvents once they enter the water column, so it makes sense to reduce the use of these toxic materials and prevent their entry into receiving waters. This in turn directs us towards preventive approaches rather than end-of-pipe treatment.

3.   What governments are responsible for the issue? Whose laws may apply?

This project was initiated by municipal governments in partnership with state and, to a lesser extent, federal agencies. Municipal ordinances were in fact one of the driving forces behind Operation: Water Works. All three levels of government could therefore be involved, but the emphasis on implementation will likely come from the municipal level rather than more "senior" governments.

4.   Who has a stake in the problem? Who should be involved in making decisions?

Stakeholders in this case include the many local businesses in the target categories, whether or not they have participated in formal pollution-prevention planning. Municipal, state, and federal governments will also be involved. Local residents and environmental non-government organizations are likely to have concerns about water quality and wildlife protection. Depending on the specific pollutants identified as problematic, medical and legal professionals may also have an interest in the problem.

5.   In the view of your decision-making group, what are the attributes of a satisfactory solution? In other words, when will you be satisfied that the problem is "solved"?

For the members of Operation: Water Works, a satisfactory solution would be cooperative and constructive, would support an ethic of water/environmental stewardship, and would be concerned with surface water quality and hazardous material. These objectives were fairly vague, and indeed impaired the project team's ability to judge its success in meeting project goals.

6.   How will you evaluate (test, compare) potential solutions?

This case suggests a program evaluation framework which was in fact applied in Operation: Water Works. This evaluation revealed weaknesses in program design (for instance, providing the "carrot" without the "stick") that can be corrected in future programs.

7.   What are all the feasible solutions to the problem?

There are many ways to attack the problem of toxic discharges to waterways. The TIP box, below, suggests some alternative approaches. The project team elected to seek "best manage-

 The Operation: Water Works Steering Committee set specific goals for the project and authorized a number of activities that were viewed as important for project success. In evaluating those activities after the fact, project staff identified a number of areas where changes could have been made to improve effectiveness. And staff also faced the problem of not being able to evaluate the ultimate environmental impact of their actions. A different project could have set different goals (either more or less specific) and undertaken different project activities. The right mix of goals and activities will depend on the wishes of the community, reflected in this case in the Steering Committee and the Business Advisory Committee, and local environmental conditions.

ment practices" on an industry-by-industry basis, then combine all the "bests" into a single plan. This is a good, simplifying, approach to a complex problem, but it does assume more homogeneity among businesses of the same type than may actually occur. A business-by-business approach is much more costly but could also be much more effective.

8.   Which solutions work "best" in terms of the attributes you identified in (5)?

To answer this question, consider the results of program evaluation presented in Section 4. This case is actually an after-the-fact analysis, which can nevertheless be very useful to us in designing projects for the future.

9.   Which solution will be easiest to implement?

Operation: Water Works staff believed that "buy-in" by business was essential to program success. The easiest approach to implement will therefore likely be the approach with widest support in the business community. This support may be (and in the view of Operation: Water Works staff, was) more important than acceptance by non-government organizations or residents. Approval by municipal and state governments is also essential for implementation, because of the legal framework within which the project will operate.

10.   What steps are needed for successful implementation? Who will pay? Who will monitor progress?

Here again, Operation: Water Works results give us insight into what is needed for successful implementation. For example, a good implementation plan might recommend hiring a public education coordinator and writer at the outset of the project, not just on an as-needed basis. Similarly, follow-up visits were requested by local businesses who had completed the planning exercise and a stronger regulatory "stick" might have improved implementation of recommended actions.

6. *Where Can I Learn More About the Ecosystem, People, and Culture of Washington State?*

The following materials provide useful background on water pollution abatement strategies, pollution prevention, and the Puget Sound area.

J. J. Breen and M. J. Dellarco. 1992. Pollution prevention in industrial processes: the role of process analytical chemistry. Proceedings, American Chemical Society 201st meeting (1991), Atlanta, Georgia.

J. C. Ebbert and K. L. Payne. 1984. The quality of water in the principal aquifers of southwestern Washington. Water resources investigations report 84:4093. U.S. Department of Interior, Geological Survey, Lakewood, Colorado.

D. T. Kollatsch. 1992. Total discharges taken into account for comprehensive planning of urban drainage and waste water treatment. *Water Science and Technology* 26(9-11): 2609-2612.

T. M. Leschine. 1990. Setting the agenda for estuarine water quality management: lessons from Puget Sound. *Ocean and Shoreline Management* 13(3-4): 295-313.

I. J. Licis, H. S. Skovronek, and M. Drabkin. 1992. Industrial pollution prevention opportunities for the 1990s. Risk Reduction Engineering Laboratory, Office of Research and Development, U.S. Environmental Protection Agency, Cincinnati, Ohio.

N. McKay. 1991. Environmental management of the Puget Sound. *Marine Pollution Bulletin* 23: 509-512.

E. K. Needham and E. L. Lanzer. 1993. The Puget Sound Environmental Atlas Update: context and GIS database integration issues. *Computers, Environment and Urban Systems* 17(5): 409-424.

G. L. Turney. 1986. Quality of ground water in the Puget Sound Region, Washington, 1981. U.S. Department of the Interior, Geological Survey, Lakewood, Colorado.

D. T. Wigglesworth. 1993. *Pollution Prevention: A Practical Guide for State and Local Government.* Lewis Publishers, Boca Raton.

New Zealand

"*How can we balance conflicting land uses in the South Coast region of New Zealand?*"

1. What Is the Background?

New Zealand's South Coast: a complex mixture of uses and resources

The Wellington Peninsula of New Zealand's South Coast was once entirely forested. With the coming of European settlement, however, much of that forest cover was removed for building, fuelwood, and similar purposes. By the beginning of the 20th century, 99% of the area had been logged. About 40 plant species have now disappeared from the South Coast ecosystem, while 29 of the remaining species are considered regionally and/or nationally threatened. Most of these 29 species are surviving in specialized habitat areas that may be vulnerable to quarrying or animal grazing. The area also supports colonies of two threatened species of weevil. Certain areas of regenerating forest provide important cover for migrating birds.

The coastal ecosystem is highly valued in New Zealand because of its threatened plants and insects. The entire coastal zone has been identified as nationally significant, while inland areas have been designated as regionally or locally significant. Distinct subzones of catchment vegetation (sheltered hillsides), coastal cliff vegetation, and coastal foreshore vegetation have been identified.

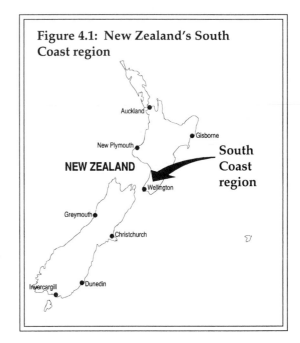

Figure 4.1: New Zealand's South Coast region

NEW ZEALAND

Auckland
Gisborne
New Plymouth
Wellington
Greymouth
Christchurch
Invercargill
Dunedin

South Coast region

Although the South Coast has been inhabited for at least 500 years, few traces of early settlement remain in the modern landscape. Yet the area contains a number of sites of spiritual and cultural value for indigenous peoples, including certain striking rock formations and rock coloration referred to in Maori and Pakeha legends. The remains of a World War II observation post are still evident on the headland, .and a number of shipwrecks have been recorded in the area, although little remains of these.

Modern land uses are diverse. Quarrying has been an important part of the local economy since the turn of the century, with most quarrying activities centered on the coastal escarpment. In the early 1960s, public outcry stopped the extension of quarrying to Red Rocks, a rock forma-

tion that figures in indigenous legends. At one time, foredune sand was also extracted from the area, but this activity has now been halted in an attempt to protect the local beach frontage.

The City of Wellington operates a landfill some distance inland from the coast, in an area called Carey's Gully. Marginal farming operations, now abandoned, have left a goat population wandering and grazing free in the area. Carey's Gully is thought to be important in providing cover for birds moving from coastal areas to areas further inland.

The South Coast area also supports a variety of recreational uses, including hiking, swimming and scuba diving, recreational boating, and fishing. Shellfish gathering is popular along some beaches. Off-road vehicles such as trail bikes currently cause noise and environmental problems, as well as conflicting with the needs of walkers and cyclists on the roads. Mountain bike riding is increasingly popular in some areas.

Box 4.1: New Zealand's environmental legislation

Resource Management Act (1991)

New Zealand's Resource Management Act, enacted in 1991, is intended to promote the sustainable management of natural and physical resources in the country. These resources are defined as including land, water, air, soil, energy, plants and animals, and structures.

Reserves Act (1977)

The Reserves Act provides for the protection of open spaces for public use and the protection of historic, cultural, biological, and archaeological values. The Act also allows for the the preservation of public access to and along coastal areas and bays and the protection and promotion of areas of special value, such as the coastal environment, the margins of lakes and rivers, and the protection of these areas from unnecessary subdivision and development.

Conservation Act (1987)

The Conservation Act is intended to promote the conservation of New Zealand's natural and historic resources, including establishment of the country's Department of Conservation. The Act requires that each regional conservancy develop a Conservation Management Strategy.

---

2. *What Problem Are We Trying to Solve?*

Finding a balance

The diverse features and resources of the South Coast ecosystem give that area tremendous potential for recreation, ecotourism, resource extraction, and other uses. It is not clear, however, what balance of these activities will best protect an ecosystem that is currently considered nationally significant. There is a strong likelihood that, without careful control of development and land use, environmental degradation will accelerate and fragile habitats and species will be lost. Furthermore, some activities may simply be incompatible, so a decision may have to be made as to which activities are allowed to continue and which are not. Finally, like many other sensitive ecosystems around the world, management of the South Coast is complicated by a mixture of public and private land ownership and the need to integrate local land-use planning with regional and national objectives.

Ultimately, this situation requires a decision to be made about an appropriate and sustainable balance of activities for the South Coast area. That decision can then form the basis for formal regulatory or nonregulatory controls on existing and future land use. Such a decision is inherently difficult because of the diversity of current activities, the complex land ownership, and the number of different stakeholders.

Who should decide what controls are needed?

The environmental and land-use tensions of the South Coast area are typical of many developed regions around the world. These tensions affect everyone in the region and also at the national level. Who should make decisions regarding future development and environmental protection?

The question of whose job it is to make these decisions is not an easy one. The New Zealand Department of Conservation has a mandate to conserve New Zealand's natural and historic resources, yet the City of Wellington probably has a more immediate and direct interest in the outcome of planning activities. In most countries, the division of power—"jurisdiction"—is usually laid out in a constitution. Often, although not always, federal (national) governments have broad powers to oversee activities that cross state or provincial boundaries—activities that are of national concern. Provinces and states may be given control over their business activities (and the pollution from them) , and sometimes over their natural resources such as forests and fisheries. Local governments, in this case the City of Wellington, are often responsible for land-use planning at the local level. It is not uncommon for several agencies, and several levels of government, to have jurisdiction—to be legitimately involved—in a single environmental issue. In fact, sorting out a process for decision-making can be a major challenge if several levels of government want to participate. Sometimes, the agency with the most central involvement takes a lead role in initiating a decision-making process and inviting other stakeholders to take part.

Should only government be involved in making environmental decisions? Regulatory agencies in most developed countries are now consciously incorporating input from the community in their decision-making processes. After all, decisions in an area like the South Coast have the potential to affect many people, including local landowners, recreational users, commercial fishers and shell-collectors, quarries and their workers, and tourists. A locally based process may be most effective in capturing local values and priorities, especially if it can incorporate a regional and/or national context for the local decision.

There is no single "correct" approach to structuring an effective multistakeholder decision-making process. A guiding rule is that everyone who has a clear interest in the outcome of the process should have a say in the decision. Some people advocate a "by invitation only" approach, by which key stakeholders are identified by the lead agency and invited to participate. Other people believe that a completely open process better serves the needs of the community by allowing anyone to become involved. Practically speaking, it is usually easier to structure a

> *Stakeholder: In environmental parlance, someone who has a direct interest in, or is directly affected by, an environmental problem.*

process that has a small, well-informed group of decision makers.

Structuring the South Coast process

In the early 1990s, the New Zealand Department of Conservation issued a recommendation for protection of the South Coast area, reinforcing widespread local concern among the community, the city government, and the media. In 1991, Wellington City Council, in conjunction with the Department of Conservation and the Royal Forest and Bird Protection Society, released a discussion document outlining key environmental and land-use concerns in the South Coast. The City Council then invited comment from all interested parties on the issues discussed in the document. A Community Advisory Committee, representative of diverse interests, was established to inform the decision-making process. The membership of the Committee was determined by nominations from the community.

An important part of the South Coast process is equitable access to information. One of the Community Advisory Committee's first activities was a field trip around the area. This activity served both as an important information gathering opportunity and as a means of building the sense of team strength that is important in effective decision making. Community Advisory Committee members were involved early in the process—in fact, at the stage of defining the problem—and remain meaningfully involved throughout. Members representing a variety of stakeholder groups explore the area's complex technical, social, and cultural issues together, developing consensus views that will be critical in implementing the preferred management strategy.

What kind of controls?

If our policy goal is to coordinate land uses so that valued social, cultural, economic, and recreational opportunities can co-exist, we can structure that coordination in two ways. One way is to write new laws that forbid certain types of activities in the area or limit the proportion of such activities in the overall mix. The other way is to develop a nonregulatory, or voluntary, framework that has the support of the stakeholders. Each approach has advantages and disadvantages; neither will work well without support from those whose job it is to implement the strategy. If new laws are written, but the regulatory agency is unwilling or unable to enforce those laws, the strategy cannot be effective. If, on the other hand, a voluntary planning framework is endorsed by city council without the support of the larger community, people will not take it seriously.

In 1994, the New Zealand Department of Conservation endorsed a nonregulatory

---

Box 4.2: Defining key legal terms

Statute: *A binding rule or law. A statute may contain wording that allocates responsibility over certain activities without providing the detailed instructions as to how that activity is to be carried out. Statutes are normally subject to scrutiny by elected representatives from all parties and can therefore be difficult and time-consuming to change.*

Regulation: *The detailed instructions appended to a statute. A regulation often contains numerical standards to be met by a particular type of activity (e.g., an industrial discharger, an agricultural operation, a quarry). Regulations are developed by bureaucrats as an administrative tool, so although they are legally binding, they are easier to change than statutes.*

Guideline: *A target or objective value that has no weight under the law. Guidelines are the least binding but also the easiest to change. They are therefore potentially the most flexible management tool, and can be easily adapted to local conditions.*

(that is, not legally binding) approach to the development of land-use planning guidelines for the South Coast. The process of developing guidelines involves significant consultation with the public, especially at the local level, with a Community Advisory Committee the focus of that consultation. Other structures would have been feasible, but this approach is certainly typical of many effective multistakeholder decision-making processes. It is intended that, once the current management planning activity has been completed, the committee will become a trust to continue oversight of land use planning activities in the South Coast area.

(Several draft management policies exist for the South Coast. These include the Draft Wellington Regional Policy Statement, indicating areas of national, regional and local significance; the Draft Regional Coastal Plan for the Wellington Region, which describes issues and mechanisms of foreshore and seabed disturbance and effluent discharges; and the Wellington City Council Draft District plan, which is a set of specific planning guidelines and areas of particular value for the Wellington area.)

3. What Components of the Environment Are Affected, and How?

Landforms and geology

The area of the South Coast of most immediate concern for land-use planning is an area extending about 8 km along the cost, between Owhiro Bay and Terawhitl Station (see Figure 4.2). This stretch of shoreline is characterized by steep cliffs rising 200 to 500 m within 300 m of

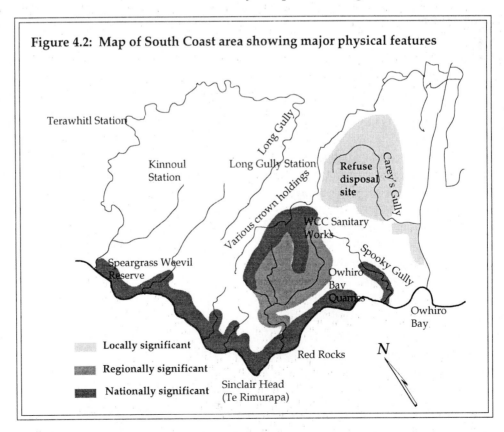

Figure 4.2: Map of South Coast area showing major physical features

the shore. There are striking rock formations resulting from ancient basaltic lava intrusions into earlier sedimentary deposits. The red pillow lava of Red Rocks is such an intrusion.

The steep slopes in much of the South Coast area support soils that are thin and erosion-prone.

Climate

The South Coast area experiences high winds relative to other coastal areas in New Zealand, as a result of funnelling of winds through Cook Strait. Annual rainfall averages 1,000 to 1,250 mm, with gale force winds occurring about 43 days each year. These conditions, coupled with an abundance of sunlight hours, create conditions that encourage high evaporation rates.

Wildlife

No comprehensive survey of wildlife has been conducted in the South Coast area. Little detailed information is therefore available for terrestrial, or marine/aquatic biota. Many unusual species are, however, known to inhabit the South Coast area. These include New Zealand fur seals, who establish non-breeding haul-out colonies on the shoreline and on offshore rocks. The resting seals may be subject to disturbance by visitors and dogs.

Other species include a rare flightless speargrass weevil (whose numbers are currently estimated at less than 200) and another, smaller, weevil. The cicada, which occurs only in isolated populations throughout the North Island of New Zealand, is found in the South Coast area.

Many bird species are found along the coastal area and inland, including a number of gull species, blue penguin, and reef heron. There is suitable breeding habitat for the blue penguin and black-backed gull along the coast, but other species are thought to nest elsewhere.

Vegetation

As described earlier, the entire Wellington Peninsula was once covered with forest, Even steep and windblown cliff areas once supported scrub forest. Much of this vegation was cleared by European settlers before the turn of the century; some areas are now regenerating.

A total of 375 species have been documented in the Wellington Peninsula. Of these, about 159 indigenous species occur in upland catchment areas. A relatively high proportion of species (compared to other New Zealand habitats) are considered threatened, mostly because of diminishing numbers and/or habitat destruction. The wide range of habitat types, even within the foreshore area, allows the coexistence of a large number of plant and animal species.

Although several ecological reserves have been established in the South Coast area, certain kinds of vegetation, such as coastal plant communities, are poorly represented in the reserves.

Generally speaking, the South Coast area contains remnants of original vegetation (which may allow further regeneration if protected), introduced species, and a variety of disturbed (human-influenced) habitats. Certain species and habitat types are considered nationally, regionally, or locally threatened and therefore are highly valued.

Sites of cultural and spiritual value

The South Coast contains a number of sites of cultural or spiritual value. The dramatic landforms of Red Rocks are mentioned in Maori and Pakeha legends, while limited archeological remains document the existence of early indigenous settlements. More recent sites include those associated with early European settlement, an army observation post, and shipwrecks along the coast. Here again, it is unlikely that a comprehensive inventory of valued sites exists, so other sites, as yet undocumented, may well exist within the study area.

Current land uses

Owhiro Bay Quarry

Quarrying has been an important commercial activity in the South Coast region for almost 100 years. Over the past 30 years, public concern about the impacts of quarrying has increased, with the effect that quarrying activities are now limited by a government-imposed resource consent, or set of rules. Quarrying at Red Rocks, and associated stockpiling of gravel on beach areas, were of particular concern.

In the late 1970s, Owhiro Bay Quarry began to explore the possibility of a land exchange, which would allow them to conduct quarrying activities in a more distant location, while abandoning quarrying at the more sensitive Red Rocks location. In 1990, the land transfer was completed, and Owhiro Bay Quarry now operates a quarry at a Wellington City Council–owned block of land farther inland. The company has permission to operate this new quarry for a 50-year period, provided that management plans are submitted and approved by the City Engineer at 5-year intervals. When these approvals were granted, the biological significance of the new quarry area, located near Spooky Gully, was not known. The company has recently offered to take actions to protect sensitive flora and fauna in the area.

Wellington City Council refuse disposal site

Wellington City Council operates a 940 hectare refuse disposal (landfill) site some distance inland from the coast (see Figure 4.2). The landfill began operation in 1976 and is expected to have a life of up to 100 years at current filling rates. The area near the landfill previously supported marginal farming operations, now defunct, and goats from those operations still roam and breed in the area.

Recreational uses

The South Coast area supports a number of different recreational uses. The particular mix of activities varies with the stretch of coastline. Along the urban coast, the most popular activities are walking, swimming, fishing from shore, skin- and scuba-diving, and recreational boating. West of Red Rocks, shellfish gathering and offshore fishing are also important recreational uses.

The coast road is popular with owners of off-road vehicles such as (motorized) trail bikes and four-wheel-drive automobiles. Vehicular traffic along this road poses safety risks and diminishes aesthetic enjoyment for walkers and cyclists. The environmental damage caused by off-road vehicles, and the noise associated with them, have been continuing concerns within the Wellington city government and with the public.

Other occasional recreational uses in the area include horseback riding, mountain bike riding, and use of beach huts (baches), some of which are deteriorating in physical condition.

Reserve areas

The Reserves Act of 1977 provides for the preservation of lands for public use, including protection of spaces that have historic, cultural, biological or archeological importance. The act is also intended to ensure continued public access to natural areas such as the seacoast. The act requires that public agencies prepare management plans for each reserve area. Currently, there are five reserve areas in the South Coast region. The Red Rocks Scientific Reserve protects red pillow lava deposits that are nationally rare. The Red Rocks Recreation Reserve, consisting of land that was formerly the site of quarrying activities, is zoned for "open space" use and has four-wheel-drive tracks. The Sinclair Head Scientific Reserve protects seal haul-out areas along the beach and on offshore rocks. The Sinclair Head Recreation Reserve was originally Maori Reserve, later transferred to the military for use as an observation post in World War II. Finally, the Speargrass Weevil Wildlife Reserve protects a small area of habitat formerly, although not presently, known to support the threatened speargrass weevil.

Other local reserves include the Karori Reservoir Reserve, which supported native bush and some plantation areas with introduced species, and there are proposals to establish reserves in catchment and marine areas. The indigenous Te Atiawa have an interest in managing the fisheries along a stretch of the South Coast.

 # 4. *How Can I Analyze This Information?*

Conflict resolution (alternative dispute resolution)

It is apparent from the foregoing sections that there are many different activities in the South Coast area, some of them likely to be incompatible. One mechanism that has been used to resolve complex environmental disputes, like the South Coast planning problem, is conflict resolution (sometimes called "alternative dispute resolution," or ADR).

Conflict resolution is predicated on the idea that in every environmental problem, there is agreement on 90% of the issue. For example, everyone wants a clean environment and a healthy economy. The debate therefore usually centers on the remaining 10%.

> *Conflict resolution is predicated on the idea that in most environmental disputes, we agree on 90% of the issue. By finding each participant's true interest in the matter, it is often possible to find an alternative resolution that meets everyone's needs.*

Successful conflict resolution demands two things. First, there must be a strong reason for everyone to stay in the discussion. Sometimes this reason can be that if consensus is not achieved, another external body will take action, for instance in the form of a regulation, that may not be satisfactory to those involved in the decision. The other important element is that the people engaged in the conflict resolution process must have the power to make decisions on behalf of their constituencies. In other words, they must be empowered to make decisions, even tough decisions, without approval from some higher body.

In all environmental disputes, each party has at heart a true concern. For a variety of reasons, an individual may be unwilling to reveal this true concern to the larger group. Instead, he or she may mask the basic interest with other issues. For example, in a search for a landfill site, it is often concern for property value that underlies stated concerns about environmental quality. *In a conflict resolution exercise, it is therefore essential that each participant come to understand the true concerns of each other participant.* Only by doing so can alternative resolutions, that meet the needs of all stakeholders, be developed.

Each participant also has a "fallback plan"—a plan that will be put into action should the conflict resolution exercise fail. If a government agency is one of the participants, its fallback plan may be to impose a regulation without consultation. If an industry is a participant, its fallback plan may be to close down—creating local job losses and related economic impacts—and move to another location. *It is therefore also essential that each participant understand each other participant's fallback plan.*

Conflict resolution is therefore essentially a three-part process. First, there is an information-gathering process, through which participants try to determine each other's true interests and fallback positions. Second, there is a stage of developing alternative approaches to solving the problem and testing those solutions out on the various stakeholder groups. Finally, there is a stage of consensus—general agreement—and unanimous acceptance, often indicated by formal signing-off.

A typical process works like this:

1. Some impartial or lead agency calls an initial meeting of the stakeholder groups.

2. At this meeting, each party is encouraged to voice his or her key issues and preferred outcomes.

3. Following the meeting, and over some specified length of time (days or weeks), individuals are encouraged to contact each other on a one-to-one basis. (At this stage, it is important that contact be just between individuals, as part of the trust-building process.)

4. At some intermediate stage, there may be a second meeting to exchange information, clarify objectives, and so on. It is not appropriate that any tentative agreements be revealed at this stage—to do so could betray the trust of fellow participants.

5. There then follows a stage of consultation between small groups of individuals. Gradually, as consensus is built, these groups should begin to coalesce until finally the single original group remains.

6. When consensus has been reached, some individual is appointed to record the group's agreement before the group (often on a flip chart or blackboard). This must be done diplomatically, seeking approval from the group for every point. It must not appear to be dictatorial.

7. In the final stage, each participant signs the agreement to indicate acceptance. The final agreement should bear the signatures of all participants.

Conflict resolution does not work for every issue or every group of stakeholders. Still, it can be an extremely powerful tool in resolving complex disputes such as exist in New Zealand's South Coast area. To be successful, the process must provide strong incentives to reach a consensus, and there must be trust and respect built among the participants. In the building of trust, it should be possible to determine each party's true interest and to find ways of accommodating that true interest in the resolution of the problem. If the conflict resolution process fails, each party will fall back to a predetermined strategy, which should also have been made clear to the participants during the process. Often, those fallback plans include more unilateral action by a government agency, which will result in a resolution of the problem, albeit one which may not meet all stakeholder's needs.

5. *How Can I Use My Findings to Reach a Solution?*

Although this case involves a number of different land uses and thus a variety of stakeholders, its resolution may be more straightforward than many environmental issues. The focus here will likely be more on the decision-making *process* than on the details that are agreed to.

1. *What is the problem?*

In Section 2, we identified the problem as finding an appropriate and sustainable balance of activities for the South Coast area, as a basis for formal regulatory or nonregulatory controls on existing and future land use.

2. *In what ways do human activities have impact on the natural environment to cause "a problem"? How do these mechanisms give you clues to possible solutions?*

The various ways in which humans impact the South Coast ecosystem are outlined in Section 3. It is clear from this description that some activities are potentially incompatible. For instance, quarrying is incompatible with protection of the heritage resource of Red Rocks. A good solution should therefore identify incompatible uses and make decisions about which of the incompatible uses will prevail.

3. *What governments are responsible for the issue? Whose laws may apply?*

The municipal level of government (Wellington City Council, for example) has a strong interest in this problem as part of its land-use-planning activities. Federal legislation is important in regulating various conservation and other activities, so the federal government must also be involved. It appears that most responsibility for environmental protection lies at the federal level, while land-use planning is a local (municipal) responsibility. This is typical of legislative roles in many developed countries.

4. *Who has a stake in the problem? Who should be involved in making decisions?*

The many different users of the South Coast are obvious stakeholders and they may bring with them non-government organizations such as scuba-diving clubs, fishing industry associations, hiking clubs, and so on. Various government agencies also have an important role, as do certain industrial interests, particularly the quarry and possibly local agricultural operations. Environmental non-government organizations would have a keen interest in the protection of South Coast species and habitats.

5. *In the view of your decision-making group, what are the attributes of a satisfactory solution? In other words, when will you be satisfied that the problem is "solved"?*

In this case, we are seeking a balance of activities—in fact, our primary goal is agreement among the stakeholders on what that balance should be. A satisfactory solution may simply be one that all the stakeholders agree to—regardless of its substance.

6. *How will you evaluate (test, compare) potential solutions?*

There is no obvious analytical methodology that can be used to compare alternative land-use strategies for the South Coast. Rather, this case involves people—having people come to a workable agreement on what they and their community believes is an acceptable and sustainable mixture of uses. In this case, conflict resolution is proposed as a method to reach consensus among diverse stakeholders.

7. *What are all the feasible solutions to the problem?*

There are probably an infinite number of land-use configurations for the South Coast. A good way to "brainstorm" some possibilities would be through a role-play exercise in which key stakeholder perspectives are represented by different individuals.

8. *Which solutions work "best" in terms of the attributes you identified in (5)?*

If the role play suggested in step 7 is employed, it can be extended to a full conflict resolution exercise. Such an exercise would allow a consensus to emerge. It's interesting to note that when the exercise is repeated with different groups of people or with the same group at a different time, a different resolution will result. In other words, there isn't a single "right" solution for this problem—what's "right" will be what serves the needs of the various stakeholders at a given place and time.

9. *Which solution will be easiest to implement?*

The *only* acceptable solution here should be one that has full buy-in from the stakeholders. If several solutions have this support, then the decision-making group may choose the least expensive or the one that will take the least time to implement.

10. *What steps are needed for successful implementation? Who will pay? Who will monitor progress?*

As with most other cases in this book, the sequence of steps needed for implementation will depend on the management decisions that are made. The role-play exercise suggested above can be extended further into a discussion of implementation needs: Who can contribute what resources? What timing of implementation is appropriate? Who is responsible for which tasks? Who will monitor and enforce?

6. Where Can I Learn More About the Ecosystem, People, and Culture of Coastal New Zealand?

The following sources may be useful in understanding the South Coast region and its ecological, cultural and economic pressures.

Ariel Dinar and Edna T. Loehman. 1995. *Water Quantity/Quality Management and Conflict Resolution: Institutions, Processes, and Economic Analyses.* Praeger,Westport, Connecticut.

I. Gabites and N. Wright (ed.). 1994. Healing the South Coast. A report prepared for Wellington City Council, Dept. of Conservation (New Zealand) and Royal Forest and Bird Protection Society. Wellington City Council (Environment Division), Wellington, New Zealand.

Benjamin J. Garnier. 1958. *The Climate of New Zealand: A Geographic Survey.* E. Arnold, London, U.K.

J. Gentilli (ed.). 1971. *Climates of Australia and New Zealand.* Elsevier, Amsterdam.

Organisation for Economic Co-operation and Development. 1993. Coastal Zone Management: Integrated Policies. OECD, Paris, France.

Wellington City Council. 1994. Wellington South Coast Management Guidelines. Wellington City Council (Environment Division), New Zealand.

Saskatchewan

"How can we minimize the environmental impact of livestock farming?"

1. What Is the Background?

More people to feed, more food to produce

The human population is expected to double sometime in the next 20 to 40 years. Although some of that population currently depend primarily on plants for its food sources, a majority of people consumes some meat in their diet.

The environmental, economic, and humane issues surrounding livestock farming are immense. In agriculture, as in many other industries, recent decades have brought a move toward intensification of production—in this case, more animals produced in a smaller land area than would have been the case 50 or 100 years ago.

The evidence of this trend is to be seen across North America and elsewhere in the world. The number of livestock farms is slowly declining, while the number of animals produced is constantly on the rise. In part, this change has arisen from the introduction of new technologies that make intensive farming more practical than in the days of hand labor. It also stems from a realization that not all land is suitable for crop production. Not all crops are suitable for human consumption, either, so as pressure to "feed the world" increases, farmers turn more and more to food products.

At the same time that agriculture is experiencing major technological and economic changes, conflicts around "appropriate" land use continue to intensify. In undeveloped areas, there may be competition for shared or exclusive land use for wildlife management,

> **Figure 5.1: Map of Canada showing location of Saskatchewan**

forestry and mining, aboriginal land claims, recreation, industry and urban development, as well as agriculture.

The quantity of arable and grazing land on this planet is finite. When good grazing lands or croplands are annexed for urban development, they are taken out of that pool permanently. As a result, farmers must now produce more, to feed a growing population, on a smaller land base. Other pressures on farmers include those to diversify in response to market demand for new crops and animal products, and strong global competition resulting in lower prices for traditional food products.

This case study examines animal production in rural Saskatchewan, one of Canada's three prairie provinces and one of the country's most important agricultural bases. Although Saskatchewan, and livestock production in that province, have their own suite of problems and issues, the kinds of pressures on the area and the industry are typical of agriculture all over the world.

2. What Problem Are We Trying to Solve?

Farming: an industry like—and unlike—any other

The challenge of modern agriculture is to provide food for a growing population without compromising the resource base that makes agriculture possible. This problem of sustainability takes many forms, including environmental sustainability, economic sustainability and competitiveness, and animal welfare. Farming is a business, not a public service. It requires high capital investment and hard work and depends on the vagaries of regional, national, and international competition for its profitability.

In recent decades, many smaller, less profitable farming operations have been swallowed up by larger, more efficient operators. Modern farms now tend to be 400% larger than they were 100 years ago, and there are only about one-tenth as many of them. As this transition has taken place, concern has been expressed about the fate of the "family farm"—the farm that has been held and operated by one family for several generations. Is there special value in this family farm unit that is lost when the unit is taken over by an agricultural conglomerate? Certainly,

> *In Ontario, another of Canada's important agricultural provinces, the livestock population now produces 20 times the volume of excrement that the human population produces—without the sewage treatment that human wastes receive.*

very large and intensive livestock operations have the potential to impact the environment much more acutely than smaller farms. In intensive livestock farming, a large number of animals may be housed under a single roof, with their feed and manures therefore concentrated in a very small area. Very large volumes of manure are therefore produced in a small area and, if not correctly managed, can have a dramatic effect on local soil, air, and water conditions.

There are several "problems" related to livestock production in agriculture. These include:

Production issues

Production issues include the need to increase production, reduce costs, remain competitive, introduce (or avoid!) new products, and generally keep pace with the industry in the face of changing weather, global economic forces, market pressures, crop failures, and product

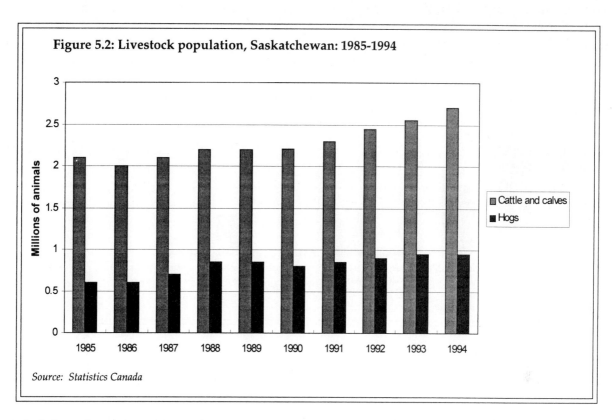

Figure 5.2: Livestock population, Saskatchewan: 1985-1994

Source: Statistics Canada

perishability. Since farming is a business, not a public right, farms that are most competitive, most profitable, and least vulnerable to weather and price changes will succeed, while more marginal operations will gradually drop out of the market.

Humane issues

Issues of animal welfare have become very important in recent years. As in Europe and elsewhere in the world, Canadian producers and the public are acutely aware of the need to treat animals humanely in rearing, transporting, and processing. In Canada, animal welfare is overseen by the Food Production and Inspection Branch (FPIB) of Agriculture and Agri-Food Canada, who administer 14 acts and associated regulations, and 11 commodity-based programs. Canada has an excellent reputation for producing safe, high-quality food, and the agribusiness sector is very aware of the sensitivity of export markets to this reputation. The FPIB monitors food exports and imports at over 200 ports of entry in Canada, but much of this monitoring is for food quality and plant and animal pests and diseases.

Environmental issues

Most farmers are at heart also conservationists, whether because of commitment to a family asset—the land—or simply because of the need to protect the land resource to preserve future profitability. Nevertheless, the environmental impacts of agriculture are still strongly debated in government, in public interest groups, and in the farming community. With respect to livestock production, issues that have been raised include:

1.   Methane releases from livestock (estimated at about 0.04% of total estimated methane releases to the environment)

2.   Odor problems from intensive swine, poultry, beef and dairy operations

3.   Transmission of diseases from wild to domestic species, and from livestock to humans

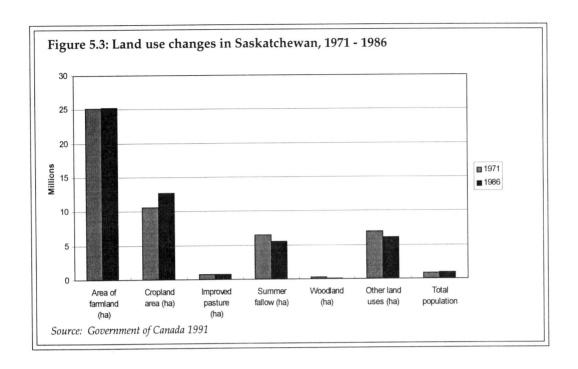

Figure 5.3: Land use changes in Saskatchewan, 1971 - 1986

Source: Government of Canada 1991

4.   Range management for the sustainable use of grazing lands

5.   Safe manure management to reduce or eliminate emissions of bacteria and nutrients to receiving waters

Residues

Modern livestock-rearing practices often involve the use of drugs such as antibiotics (to treat illness) and production stimulants such as steroids. Human consumers are understandably concerned about the potential for residues of these materials to remain in their food. In Canada, there is close scrutiny by the federal government over food quality, both food for domestic consumption and food for export. Safeguards and testing are built into the system at several steps. For instance, a food packer is under no obligation to accept animals from producers who have been known to violate food safety requirements, so a careless livestock producer may have trouble selling his or her products.

Microbiological residues are more difficult to tackle. Virtually all meat products contain some bacteria, and even with the best technologies in dressing, packing, and storing meat, and with careful inspection, bacteriological contamination is a frequent problem. Technologies such as food irradiation are available but have been slow to gain acceptance by the public because of their own health and environmental implications.

Animal welfare, food quality, and residue issues are largely a matter of education, regulatory, or other controls and careful inspection and enforcment. For the environmental manager, however, the air, water, and solid waste impacts of livestock rearing can be a sufficiently complex challenge. For the purposes of this case, we will focus on the environmental impacts of livestock farming and certain related issues of animal welfare. The problem therefore becomes how to identify potential sources of impact, how to develop suitable approaches for control or prevention, and how to convince farmers to implement those approaches once they have been identified.

3. What Components of the Environment Are Affected, and How?

Saskatchewan's agricultural base

In 1995, Saskatchewan had a farmland base of about 26.6 million ha, of which about 21 million was cultivated. Of the 21 million, 13,300,000 is land on which crops are grown, 2,700,000 is pasture and hay, and 5,300,000 is native grassland, or rangeland as it is sometimes called.

Farming this land are 60,840 farms with an average size of about 436 ha, operated by 159,700 people. Saskatchewan's total 1995 population was 988,400, of which 365,000 lived in rural areas, so the farming population makes up a little less than half the total rural population of the province.

In 1995, there were approximately 2,700,000 cattle and calves in Saskatchewan, while the pig population, at about 900,000, is about the same size as the human population. Other livestock reared in the province include horses (about 8,000) and a variety of species on game farms: 2,800 buffalo, 5,500 elk, 2,800 deer and 1,000 wild boar. Various exotic species such as ostrich are also raised, although in much smaller numbers.

Cattle production

Cattle breeding

The cattle now raised in Saskatchewan represent a variety—and a mixture—of breeds including traditional breeds and more exotic breeds introduced in the 1960s and 1970s. Breeds are not always maintained "pure" but may be cross-bred to achieve desired changes in frame size, lean-to-fat ratio, growth rate, and ultimate size. Although this crossbreeding achieved "heterosis," or hybrid vigor, it also created a number of genetic, fertility, calving, and temperament problems. Over the past three decades, these problems have largely been resolved, but certain problems (and potential problems) may remain.

Livestock farmers look at their operations as businesses, selecting the best offspring as potential replacements for brood stock.

> *Registered stock: Cattle of known genetics, attractive to the commercial industry and other breeders because of their predictable growth patterns, appearance, size and yield.*

Feedlot operation

The number of cattle owned by a single producer varies widely, from a few hundred animals ("head") to many thousands. As profitability margins decrease, economies of scale have encouraged producers to move to larger operations, allowing them to spread overhead costs such as barn construction and heating across more animals, and also guaranteeing a steadier and more consistent supply to the marketplace.

Many cattle operations raise animals in fenced enclosures called feedlots. Although this decreases the amount of land needed for free grazing, it also concentrates wastes and odors in a small area. In intensive operations, and those raising breeds which are sensitive to heat and cold, the animals may spend much of their lives inside buildings. Animals confined in this way can be subject to aggressive behavior, so they are often housed in small groups to reduce the risk of injury through fighting. This type of livestock enclosure is designed for centralized collection of manures, usually through floor grates that divert wastes away from the floor through collection pipes to storage facilities. Animals raised in intensive, enclosed operations with controlled feed frequently have more liquid manure than those living in the open. It is essential that these liquid wastes be stored properly and safely, away from receiving waters, and that they be capped and vented to avoid odor problems or risks of the explosive or poisonous gases that can form in anaerobic storage conditions.

When many animals are raised in a small space, the risk of disease, fungal, and parasitic infection increases. Some breeds are especially susceptible to certain diseases. These concerns have led to increased use of antibiotics and other pharmaceuticals in modern intensive livestock operations.

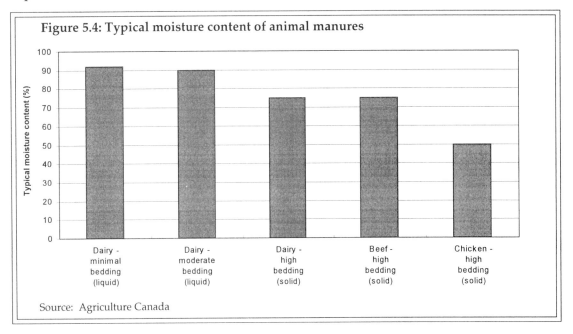

Figure 5.4: Typical moisture content of animal manures

Source: Agriculture Canada

Meat processing and packing

The economics of meat packing require a steady stream of animals to be delivered to the meat-packing plant. This allows steady line speeds to be maintained, a constant level of staff to be employed, and storage facilities to be used efficiently without waste or excess. Livestock are sold at open auction, at terminal markets in large urban centers (where an agent exacts a commission for arranging a sale between producer and buyer), and at direct markets (where a sale is agreed to between the producer and buyer without an agent's help). Direct markets can be risky for novice producers, who may not get the best price for their stock in this way. Meat packing is usually a separate operation from livestock rearing, but some meat packers also operate feedlots.

Although meat is the primary focus of livestock marketing, the "drop"—the nonmeat components of the animal, including hide, feet, edible and inedible viscera, and by-products—can account for as much revenue to the packer as the dressed meat itself. These components are used to produce textiles, brushes, leather goods, buttons, glue, knives, tallow (for use in shortening and soap) and drugs such as epinephrine and insulin.

When an animal reaches the meat-packing plant, it is first stunned using mechanical or electrical stunners, then it is killed and the hide and internal organs removed ("dressing"), then the carcass is cut into halves, washed, and refrigerated. Chilled and graded meat is then packed as boxed or "block-ready" beef (sides and quarters) and sent out to retailers.

4. *How Can I Analyze This Information?*

Identify the sources

What pollutant sources are created in livestock management? As a first step in assessing the situation, determine the nature, volume, and quality of these sources. Some likely candidates are:

Spilled and leaked manure

Incorrectly stored manures (mixtures of urine, feces, and bedding materials) can spill or leak from storage facilities or from uncontained storage in exercise yards and feedlots. The consistency and chemical composition of manure varies considerably with breed, sex, size, feed, and confinement. Some manures are intentionally diluted with water before storage.

Contaminated runoff

Rain and melting snow draining from feedlots and barn areas can pick up contaminants from open piles of solid manure. Other sources of contamination of runoff include silo seepage, spilled chemicals and fuels, and similar materials.

Livestock housing wash water

Where livestock are confined in a building, water used to wash animals and building surfaces can become contaminated with bacteria and nutrients.

Milk house wash water

Dairy operations use large quantities of water in washing surfaces and equipment for milk production. This wash water is rich in phosphorus and other contaminants.

Identify appropriate remedial and preventive measures

Although odor is a constant problem with stored manure, more serious environmental problems arise when flowing water comes into contact with stored or open manures or milk wastes. This water can enter lakes and streams in many ways, including direct seepage, rainfall or snowmelt runoff from feedlots, and runoff from fields where manure has been spread as a fertilizer. In choosing remedial and preventive measures, therefore a basic principle is to minimize the contact of water with wastes.

Store manure properly

Whether the manure is liquid or solid, appropriate storage of adequate capacity is usually the first step in correcting a livestock pollution problem. The storage facility should be large enough to store all manure, bedding, waste feed, and any associated liquids until the farmer can treat or dispose of it. In Canada, it is common to apply stored manures to farmland as a fertilizer. Ideally, these wastes should be spread and then incorporated into the soil immediately after spreading. Application of manure in wet weather, or in the winter on frozen ground that will later thaw and drain, increases the likelihood of groundwater and stream contamination. Adequate manure storage capacity allows the farmer to delay application until the ground is dry and the risk of contaminated runoff is reduced. At least seven months of storage capacity is advisable, eight if the soil is naturally wet and poorly drained.

Solid manure can be stored with bedding materials in the barn (which may limit the available storage capacity), in separate roofed storage, or on an open concrete pad having a drainage system. Liquid manure is stored in covered or open tanks or earthen ponds. Covered tanks can be situated above ground or in-ground, even beneath the barn floor so, that wastes drain directly into them.

Divert rain and snow

A second step in preventing pollution from livestock operations is to keep uncontaminated water such as rainwater separate from animal wastes. Useful approaches include installing eavestroughs on structures, covering or roofing open manure piles, and berming (a berm is a raised curb or barrier) or guttering feedlots and exercise yards to collect and divert falling rain and melting snow.

Cover storage facilities

Manure storage need not be covered, but where odors are a problem a cover is often essential. Covers for in-ground storage tanks must be able to support the weight of heavy farm machinery. All covers should be structurally sound, lockable, and well-vented to prevent the build-up of noxious gases.

Store runoff and livestock housing wash water

Runoff diverted from open areas can be stored in tanks or ponds for additional treatment (e.g., settling of suspended particles, reduction in biochemical oxygen demand) before safe discharge to a lake or stream or infiltration into groundwater. Such storage may not be necessary for uncontaminated water, but if there is any chance of contamination, or if the operation generates large quantites of wash water, it is safer to store it temporarily than to discharge it directly.

Treat or store milk house wash water

Milk house wash water is a special case of contaminated water disposal. Milk house wastes are typically higher in phosphorus and decaying organic material than are rainfall runoff or livestock housing wash water.

One option for dealing with milk house wash water is to store it with manure or runoff (obviously the sizing of the storage facility must be adjusted to accommodate the extra volume from this source). Although this is cheap and easy, it doesn't achieve the level of pollutant reduction that would be possible with a separate facility.

A better approach is to create a treatment system similar to a household septic tank that affords some anaerobic treatment followed by discharge into a weeping tile bed. Usually, the contaminated water is pumped or drained by gravity to an underground storage tank large enough to hold about a week's worth of flow. In the tank, anaerobic processes reduce contaminant concentrations, and the effluent from the tank then flows to a perforated pipe laid in crushed stone (the "weeping bed") and then into the soil. In the soil, any residual phosphorus can adsorb to clay particles or combine with calcium or iron.

Other approaches to reducing the impacts of milk house operations include rinsing milk lines with a small quantity of water that is then fed to older calves. Wash water can also be recycled and reused, reducing water consumption and wastewater volume by almost 50%.

Assess the impact of management actions

The problem of environmental impacts from livestock operations is not "solved" until we can observe improvements in the environmental characteristics that concern us. Several tools are available for this purpose.

Computer simulation models

A variety of simulation models are available in the the literature from agricultural research institutions and from government agencies concerned with agricultural activities. These tools range from simple inventory programs that allow the analyst to list potential hazards in a farm operation and assess the hazard reduction possible with a single or with several remedial measures, to sophisticated computer simulation methods generating hour-by-hour predictions.

More complex simulation tools include models (such as CREAMS) that simulate pollutant loadings from agricultural activities. This type of model can also be used to evaluate the impact of management actions on a given operation or watershed.

Water-quality simulation models such as the U.S. Environmental Protection Agency's QUAL2E, WASP5, or HSP-F allow the analyst to simulate the response of a river to changes in pollutant loading. Some, like QUAL2E, are "steady-state" models that assess pollutant distribution under a single set of environmental conditions (for example, a particular streamflow and water temperature). Others, like WASP5 and HSP-F, are "dynamic" models that reassess stream response at every model time step (usually every 15 minutes to 2 hours of model time). Both steady-state and dynamic models can be used to examine stream response at different places in the stream; the modeler can decide what is the most appropriate division of the stream into sections or "reaches" for more detailed analysis.

Site visits

Even without sophisticated computer tools, a visual survey of a farm operation can be a very useful tool for assessing pollutant sources, transport mechanisms, and likely impacts. An on-site visit also gives an opportunity to discuss with the farmer any considerations of cost, work schedule, and physical infrastructure on the farm. What does the farmer see as the operation's main strengths and weaknesses? What changes might be acceptable to the farmer? Under what conditions?

Ongoing monitoring

Finally, ongoing monitoring is a very useful tool to track the impact of current practices or the response to improvements made in the operation. Monitoring gives the farmer reassurance that scarce cash resources have in fact been put to good use, and it provides an early warning of emerging problems before they become too costly to fix.

5. How Can I Use My Findings to Reach a Solution?

Designing an environmentally-protective livestock operation is a fairly simple process using the problem-solving framework described in the Introduction.

1. *What is the problem?*

In Section 2, we identified the problem as how to identify potential sources of environmental impact from livestock operations, how to develop suitable approaches for control or prevention, and how to convince farmers to implement these approaches once they have been identified.

2. *In what ways do human activities have impact on the natural environment to cause "a problem"? How do these mechanisms give you clues to possible solutions?*

Livestock operations have impact on the natural environment because they bring large numbers of animals together, often in very small areas. As a result, animal wastes which in natural systems would be distributed over a wide area are concentrated in space and time. When a feedlot is located close to a river or lake, or when manure is applied to wet ground, there are additional opportunities for the flow of pollutants into receiving waters. Toxic gases and odors from feedlot operations can also have negative impacts on neighboring residential development. These mechanisms suggest that solutions could lie in reducing the concentrations of animals—not a very attractive option for a feedlot operator—or in reducing the potential for pollutants to enter air and water.

3. *What governments are responsible for the issue? Whose laws may apply?*

In Canada, the control of pollution from agriculture is a responsibility shared by the provincial (state) and federal governments. In many provinces, the provincial government has taken the lead in this area; in a few, the federal government has traditionally had the stronger role. Both levels of government have laws that could potentially apply to the control of pollution from agriculture.

4. *Who has a stake in the problem? Who should be involved in making decisions?*

Pollution control and pollution prevention usually cost money. Adding new manure storage or runoff containment can add to the costs of raising livestock, thus raising the price that the farmer must charge for the animal at market. Important stakeholders therefore include feedlot operators, meat packers, and meat consumers, as well as the governments listed in step 3.

5. *In the view of your decision-making group, what are the attributes of a satisfactory solution? In other words, when will you be satisfied that the problem is "solved"?*

Appropriate clean-up targets could include "best management practices" (in other words, the problem will be "solved" when every feedlot operator has covered, bermed manure storage), or ideal concentrations for pollutants of concern (e.g., total phosphorus in the receiving stream < 0.02 mg/L). State and federal government guidelines for air and water quality can be very useful in determining what is an "acceptable" level of discharge.

6. *How will you evaluate (test, compare) potential solutions?*

There are probably two important considerations here: performance (i.e., pollutant reduction) and cost. Costs can be obtained from local contractors and equipment suppliers. Pollutant reduction can be estimated in various ways. If no predictive model is available, it should still be possible to rank possible solutions (e.g., good, better, best) or to estimate their effectiveness from the scientific literature. Predictive models can provide a more detailed projection of an option's performance in space and time.

7. *What are all the feasible solutions to the problem?*

Most of the commonly-used approaches are described in this case. Don't hesitate to think up new ones or consult experts (including the scientific literature) on emerging technologies.

8. *Which solutions work "best" in terms of the attributes you identified in (5)?*

Use the methods described in step 6 to evaluate the options you identified in step 7. You may wish to weight some criteria (for instance, cost) more heavily than others. If you do assign weights, it's useful to have all the individual weights sum to 100. (As an example, you might assign a weight of 50, or 50%, to cost, 20 to nitrate reduction, 20 to phosphorus reduction, and 10 to bacterial reduction). This avoids the problem of skewing your analysis with extreme arbitrary weights (e.g., 200 for cost and 3 for nitrate). Simply multiply each option's performance (rank, percentage reduction, dollar cost, or some similar measure) on a given attribute by the weight for that attribute to arrive at a total score. The "best" option is often the one with the best overall score.

9. *Which solution will be easiest to implement?*

The success of this type of plan will depend very much on farmers' willingness to accept change. Cost is certainly a factor here, so an important implementation tool could be government grants and subsidies for clean-up measures. Farmer acceptance is so important in this type of decision that analysts will often include "implementability" in their list of decision criteria (step 5) and their analysis of performance (step 8).

10. *What steps are needed for successful implementation? Who will pay? Who will monitor progress?*

Implementation planning could be as simple as arranging a chat between local technical support personnel and an interested feedlot operator. More often, it will require federal and state agreement to provide funds for grants and subsidies, or to guarantee strict enforcement of existing pollution control legislation. Implementation planning may also involve setting priorities on the worst—most polluting—operations first, to optimize the environmental return on each dollar spent on clean-up.

6. *Where Can I Learn More About the Ecosystem, People, and Culture of Rural Saskatchewan?*

The following sources give background on livestock management in general, animal waste management issues and control, and animal welfare concerns.

Agriculture Canada and the Ontario Ministry of Agriculture and Food. (undated). Best Management Practices: Livestock and Poultry Waste Management. Agriculture Canada and Ontario Ministry of Agriculture and Food, Guelph, Ontario, Canada.

Canadian Agricultural Research Council. 1992. Proceedings of the National Workshop on Land Application of Animal Manure. Ottawa, Canada.

Government of Canada. 1991. *The State of Canada's Environment*. Queen's Printer, Ottawa, Canada.

Ken Simpson. 1986. *Fertilizers and Manures*. Longman, London, U.K.

John F. Vallentyne. 1978. *U.S.-Canadian Range Management: 1935-1977: A Selected Bibliography on Ranges, Pastures, Wildlife, Livestock, and Ranching*. Oryx Press, Phoenix.

Some of the best information available to you may come from local extension specialists from agricultural agencies. This information is often produced in factsheet or booklet form—succinct, factual, and tailored to the kind of problem we consider in this case study.

"How can we decide which of many problems in the Ganges River is most urgent?"

1. What Is the Background?

A holy river

For thousands of years, the Ganges River, revered as Mother Ganga, has supported human settlement along its banks. In modern India, the river's catchment area of 900,000 km^2 covers eight states and occupies 43% of the irrigated area of India. Its valley is densely populated in parts, with several of India's largest cities, including Calcutta (metropolitan area about 11,000,000 people), Kanpur (about 3,000,000), and the holy city of Varanasi, also known as Banaras or Benares (about 1 million).

The Ganges, as the goddess Ganga, is central to the ancient Hindu religion, a source of spiritual strength and renewal. Hindus believe that the holy river has the power to wash away sins. On a single holy day, more than a million people may bathe at a single location. The waters of the Ganges are considered to have magical powers, including the power to stay fresh for a year when bottled. Despite the fact that the river receives discharges from innumerable sewers and industrial outfalls, devout Hindus believe that the waters of the Ganges are everywhere pure and safe to drink. Apart from holy bathing, millions of Hindus use the river every year to carry away the ashes of cremated bodies, or the uncremated bodies themselves.

Figure 6.1: Map of India showing location of Varanasi and Ganges River

65

Flooding and flood damage are increasing; water pollution is suspected

The Ganges basin is urbanizing rapidly. In upland regions draining to the river, the rate of deforestation and grazing pressure have increased steadily over the past 30 years. Without vegetation to hold them, soils erode quickly. Deforested slopes quickly lose their water-holding capacity, allowing the heavy rains of the monsoon season to wash downslope into the river. Downstream, the Ganges—once a stable hydrologic regime—now floods every year, with tremendous loss of human life, property, and livestock.

Water pollution in the river is also of growing concern. Perhaps because of its holiness, water-quality data for the Ganges are scarce. The city of Varanasi (formerly Benares), in particular, is considered holy as a place of "crossing over" for the dead and dying. Yet Varanasi is also a busy industrial center known for its textile production, carpets, and various other products. Varanasi has 8 major sewage outfalls and an estimated 70 open sewers entering the river. About 30 million dead human bodies are cremated each year at two *ghats*, or bathing steps. These cremations create about 100 tonnes of ashes each month, all of which are disposed of in the river. Unburnt and half-burnt carcases of animals and humans also enter the river, because many families cannot afford the high cost of wood to fuel fires on the ghat. These materials consume oxygen as they decay, and promote the growth of bacteria and fungi in the river. Scavengers pick through the river muds for jewelry and gold fillings from the corpses.

Industrial wastewaters from large and small operations are another major source of pollutants to the river. Less important pollution sources arise from clothes-washing, bathing, defecation, and agricultural drainage along the river banks.

Box 6.1: Water pollutants observed in the Ganges River at Varanasi

Oxygen-demanding materials from human and animal excrement, industrial wastes, decaying plant and animal material, uncremated human and animal remains. Foul-smelling; reduced instream oxygen affects fish and other aquatic life.

Bacteria and viruses from human and animal excrement, uncremated human and animal remains. Contribute to the spread of waterborne diseases such as typhoid and cholera, polio, hepatitis.

Phosphorus and nitrogen from sewage and industrial wastes, agricultural drainage (fertilizer), animal manures. Nutrients for plant growth; high levels encourage algae blooms and excessive aquatic weed growth, which in turn can deplete dissolved oxygen.

Synthetic organic pollutants from industrial discharges, detergents, pesticides and herbicides, pharmaceuticals, and other sources. May have immediate (ranging from skin irritation to death depending on concentration and organism affected) or long-term (e.g. neurological impacts, cancer, miscarriage) effects on humans and other organisms.

Inorganic pollutants especially metals and metal compounds. Effects depend on compound and affected organism, as for synthetic organic pollutants, above.

Sediment from agricultural and urban runoff, soil erosion from upland areas, construction activities, ashes from cremated bodies, industrial discharges, and other sources. Impairs appearance of water; can also contribute to silting of reservoirs and channels; can affect gill function and spawning substrates for aquatic organisms.

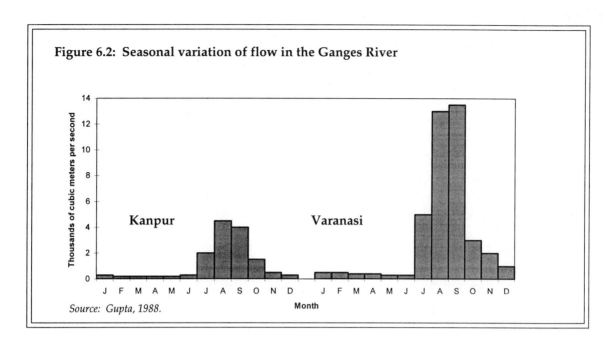

Figure 6.2: Seasonal variation of flow in the Ganges River

Kanpur

Varanasi

Source: Gupta, 1988.

The Ganga Action Plan

By 1984, dysentery was killing more than 50,000 people along the Ganges every year. Infant cholera deaths had doubled since 1980. Half of the river's large cities, including Varanasi, whose population is over a million people, had no sewage treatment. In 1985, Indian Prime Minister Rajiv Ghandi officially recognized the degraded quality of the Ganges, announcing a national program to restore the polluted river. The government promised a nation-wide effort, including education programs, youth involvement programs, sewage works, and many other programs. This Ganges Action Plan cost in excess of $425 million U.S. It has received much praise and much criticism.

The reality is probably somewhere between success and failure. The budget—huge in a country where a month's wages can be measured in tens of dollars—was in fact too small to construct treatment works in all the river's cities.

Many successful actions were undertaken. The project constructed extensive drains and sewer systems, riverside toilets, and sewage treatment plants, mostly in the 29 Ganges cities with populations over 50,000. Electric crematoria were built in holy cities like Varanasi, where fuelwood cremations had formerly been the norm. (Traditional wood-burning cremations cost about $20; the new electric crematoria cost only $2 per cremation.) In Varanasi, project staff began to encourage the breeding of native flesh-eating turtles to speed the removal of decaying carcasses in the river.

The project was, however, slow in gaining momentum. Endless bureaucratic delays and tendering debates—domestic suppliers? international technology?—meant that half-way through the 5-year plan, less than one-quarter of the budget had been spent. About 410 million of the planned 873 million liters of sewage is now diverted to treatment works, and about two-thirds of the 70 or so worst industrial polluters have now installed effluent treatment systems. Neither industrial nor municipal sewage works function during power failures, however, and in India power failures are almost a daily occurrence. Engineering blunders have resulted in clogged sewers and even sewers that lead nowhere, allowing waste to back up into streets and houses.

2. What Problem Are We Trying to Solve?

A daunting array of issues

Despite expenditures of millions of dollars and extensive capital works, the Ganges remains an area of impaired water quality, devastating floods, and intensive religious, municipal, agricultural, and industrial use. Clearly, the river still has many problems. The issue here is to decide which of them demands attention first.

This is no easy problem. How do we rank problems objectively without imposing our own preconceptions about what is bad and good, important and trivial?

Several tools are available to help structure such an analysis. The user of these tools should be aware, however, that solving a problem is not as simple as identifying it and picking a solution. Effective problem-solving also demands community involvement in all stages of problem identification, development of solutions, and implementation of a preferred approach.

For the purposes of this case study, you should assume that all interested stakeholders are contributing to the analysis. Our immediate problem, then, is to sort out the important issues from the less important. In a situation as complex as the Ganges River at Varanasi, there can appear to be an overwhelming number of issues: social, economic, population, industrial pollution, human health, aesthetics, and so on. It's easy to become confused as to which is most urgent, particularly when advocates of different viewpoints are highly vocal, or even aggressive. Media—television, radio, newspaper—coverage can also skew an observer's view of the relative importance of issues.

In a sense, the "problem" we confront in this case therefore is to decide which of the many issues before us is most urgent. The methods presented in Section 4 can be helpful in sorting out these priorities.

3. What Components of the Environment Are Affected, and How?

Identify key environmental components

Once you have an issue clearly defined, the next problem is to determine what aspects of the environment are relevant to the issue. This is not to say that a researcher should focus on a tiny aspect of the natural environment without examination of the larger ecosystem. What is important is that we apply limited resources (time, money, people) to the parts of the ecosystem that are central to the problem and its mechanisms. For example, if we are concerned about industrial discharges of certain persistent toxic chemicals, because of their potential to cause cancer in humans, it may not make sense to spend time studying the structure of forest ecosystems in upland areas, or bacterial contamination of water supplies. Instead, we would want to understand the volume, quality, and variability of industrial discharges, the routes by which those chemicals can enter the human body, and the ways that the body detoxifies or accumulates toxic materials. If we understand these issues, we can begin to think about possible solutions for the problem: discontinue or reduce discharges of toxics, reformulate products and alter processes so they use less toxics, develop pharmaceutical products that help the body detoxify toxic substances more effectively, or whatever.

How does one go about deciding which parts of an ecosystem are important or relevant to a given problem?

A first step: understanding ecosystem structure

A good starting place is to think about the structure of the ecosystem affected by the problem, and the dynamics of materials cycling and energy flow within the ecosystem. How does the harbor ecosystem behave? What plants and animals live in the ecosystem? How do those species act and interact, in terms of predator/prey relationships, habitat requirements, dependence on abiotic factors, and other similar factors? What food chains or food webs exist that are relevant to the issue you are considering?

Next, think about the cycling of materials and the flow of energy

How do materials cycle in the ecosystem? How does water move through it? How variable are these processes in space and time? What biogeochemical cycles are likely to be important here: nitrogen and phosphorus? carbon? oxygen? Would it be helpful to construct a mass balance of a particular material, such as mercury, through the ecosystem, so we know which environmental components hold which proportion of the total mass? Is biomagnification occurring? If so, for which

Keep the problem you are trying to solve clearly in mind when you consider ecosystem structure and function. If you allow yourself to be distracted by other problems, you run the risk of complicating your analysis significantly. This is because the solutions for industrial discharges of heavy metals, for example, are not the same as the solutions for aesthetic problems or the problem of high dysentery rates in children under 5. Consider one problem—one ecosystem context, one set of possible solutions—at a time.

species, and which substances? Materials tend to flow through systems in a circular manner: into organisms, back to the abiotic environment, then back to organisms. Can you map out this cycle in a general way?

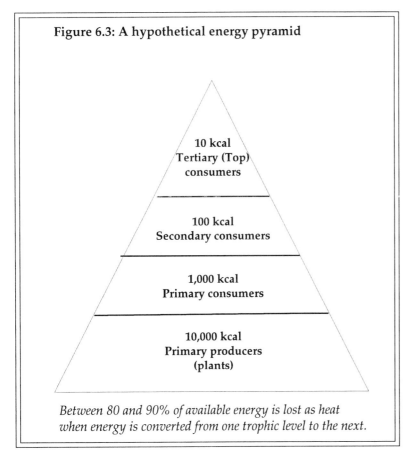

Figure 6.3: A hypothetical energy pyramid

10 kcal
Tertiary (Top)
consumers

100 kcal
Secondary consumers

1,000 kcal
Primary consumers

10,000 kcal
Primary producers
(plants)

Between 80 and 90% of available energy is lost as heat when energy is converted from one trophic level to the next.

In contrast to the cyclic flow of materials, energy flow on earth is one-directional. The first law of thermodynamics states that energy can be changed from one form to another but it cannot be created or destroyed. The second law of thermodynamics essentially states that "you can't get something for nothing"—in all energy transformations, some energy is always lost as heat. How does energy flow through the ecosystem you are interested in? Where is most of the primary production occurring? What about the secondary production? Gross versus net productivity? Can you develop a rough energy (productivity) pyramid for the ecosystem? A productivity pyramid shows the primary producers as the foundation of the pyramid, with secondary producers/ primary consumers next, then higher levels of consumers, and finally the "top carnivore" at the peak of the pyramid. The pyramid shape is derived from the inefficiency of energy conversion from one trophic level to the next: of all the energy fixed by plants in a given ecosystem, only about 10 to 20% is available to the next level for tissue building and maintenance; the rest is lost as heat.

Box 6.3: Some important abiotic factors to consider in ecosystem evaluation

• *Light (=> photosynthesis)*
• *Water (a solvent, a heat reservoir)*
• *Salinity*
• *Temperature*
• *Oxygen supply*
• *Nutrient supply*
• *Fire*
• *Soil type*
• *Movement of water and air masses*

Finally, consider the nonliving parts of the ecosystem

What about the nonliving, or "abiotic," components of the ecosystem? Keeping the issue you are trying to resolve clearly in mind, which of the factors in Box 6.3 might be particularly important in determining the nature and mechanism of the problem?

If you can answer these questions even in a general way, keeping the issue you are trying to resolve clearly in mind, a picture of the ecosystem will begin to emerge.

4. How Can I Analyze This Information?

A two-step process

In a sense, analyzing the problems of the Ganges River is a two-step process. First, we must sift through information on a variety of candidate problems to determine which is most urgent and demands attention first. Given the limited time, money, and human resources available to most decision makers, it makes sense to solve the worst problems first, then move on to less pressing issues. Some of the analytical tools available to screen problems are those presented below.

When a problem has been clearly identified, and the biophysical environment well understood, there remains the problem of evaluating the human social, cultural, ethical, and economic context of the problem and its mechanism. After all, a set of biophysical circumstances, like flooding, might be an unremarkable occurrence in one set of circumstances, but a significant hazard in another. Finally, we must develop a series of possible solutions to the problem and evaluate them to determine which is "best" given the social, cultural, and economic context, for resolving the problem we observe. Methods such as those presented in this section are also used in evaluating alternative management strategies.

Clarifying your goals

Albert Einstein once said that if he had 100 days to save the world, he would spend 99 of them clarifying the goals and a single day on action. Environmental policy is often set without a clear policy goal in mind—in other words, without an understanding of what change is ultimately desired and why. Your decisions will be clearer if you can first decide what you are trying to achieve with your action. For instance, is protecting human health your first concern? Or perhaps wildlife habitat? Species diversity? Minimizing energy usage? And so on. It will soon become clear that a policy that minimizes energy usage will not necessarily protect human health, and vice versa. Your goals should be clear and succinct. Set them early in the process, and don't stray from them without careful thought.

Risk assessment

Risk assessment is a quantitative method that assesses the *consequences* of a hazardous or adverse event, coupled with the *probability* that the event will occur. Risk assessment is often applied to problems of standard setting: how much dioxin is "safe", for example. It can also be applied to problems of resource depletion, such as loss of plant or animal species, soil erosion, overgrazing or deforestation, or environmental accident, such as nuclear accident, chemical spill, or volcanic eruption. Risk assessment is still controversial as a method of standard-setting, because people do not always agree on the fundamental com-

> *Risk assessment: A process of evaluating the consequences of an adverse or hazardous event, multiplied by the probability of that event occurring. The consequences of an event may be judged differently by different cultures.*

ponents of the analysis. For instance, determination of the consequences of chemical exposure requires that we define a time period over which a "dose" can be calculated. What is an appropriate time period? Is it a one-time exposure? A life-time exposure? And so on. Similarly, people disagree as to which information should be used in assessing toxicological response. If the only data available for a contaminant comes from laboratory studies of guinea pigs, does that data have any relevance for human health or fish toxicity?

Despite this debate, risk assessment can be a useful tool in setting priorities for a complex problem. Properly applied, the method has four major steps: identification of a risk; estimation of the consequences of the event occurring, usually in a quantitative way, through an understanding of the pathways and processes by which risks occur; and evaluation of the social and cultural context of the risk. The final step is that of policy-making and management of the risk—an action plan, in fact.

In the case of the Ganges, we might identify one risk as chromium from industrial discharges. To estimate this risk, the analyst could search the scientific literature for toxicological studies showing how different chromium compounds affect humans or other organisms. In evaluating the risk, the analyst would have to place the risk in its social and cultural context. If exposure to a certain chromium compound causes adverse health effects in an additional one in 100,000 people, how does that compare with adverse health effects from voluntary risks such as smoking and driving, or involuntary risks such as being struck by lightning or living in an area where snakes are present? It is important to realize that whereas risk estimation is a quantitative step, risk evaluation is a more qualitative process, ultimately resulting in a political "judgment call" as to what action is necessary or desirable.

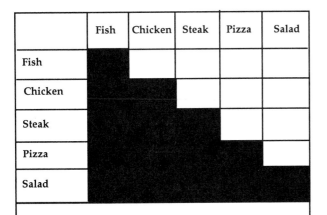

Figure 6.4: Example of pairwise comparison

	Fish	Chicken	Steak	Pizza	Salad
Fish		Chicken	Steak	Pizza	Fish
Chicken			Chicken	Chicken	Chicken
Steak				Steak	Steak
Pizza					Pizza
Salad					

Pairwise comparison

Pairwise comparison is a method of structuring decisions that can help the analyst avoid unconscious bias toward or against particular options. Much has been written on the method, but stated simply it works as follows.

The analyst identifies separate options, with the idea that only a certain number, often one, will be chosen as the preferred strategy. In a simple example, we might try to decide whether we will have fish, chicken, steak, pizza, or salad for dinner. The options are then arranged in a matrix with all options listed along the horizontal axis and again down the vertical axis (see Figure 6.4). Note that one-half of the matrix is blocked out because it is simply a mirror image of the other half.

The analyst, alone or (preferably) in consultation with a wider stakeholder group, then considers the options "pairwise" and decides which of each pair is preferred. Working through our example matrix, we could enter the preferred option in each matrix cell. The result might look something like the matrix in Figure 6.4.

It's clear from this simple analysis that the analyst prefers chicken first, then steak, then pizza, then salad. Fish didn't even make the list!

If we apply this technique to the Ganges River, we could list the various issues (for example, industrial discharges of heavy metals, decaying flesh in the river, erosion and flooding, etc.) as options for action, structure them in matrix form, then decide pairwise which of each pair is most demanding of attention. Although this decision-making method is still open to bias, like risk assessment it offers a way to assign priorities to conflicting demands.

Multiattribute comparison

Another approach to decision-making is to decide what societal goals are most important and then decide what actions are necessary to achieve those goals. For example, we might decide that protection and improvement of human health is the single most important goal we have. Secondary goals might be restoration of a fishery in the river and improvement of aesthetics.

We could then set up a matrix, this time with the various issues down the vertical axis and the goals across the top. The matrix could then be filled with ratings for each issue in terms of its impact on each goal. The ratings themselves can be qualitative (e.g., high, medium, low) or quantitative (e.g., rated on a scale from one to ten); it doesn't really matter as long as you use a consistent approach.

When all the cells in the matrix have been filled, we can then add the scores (or count the number of highs, mediums, and lows) to determine which issue is most pressing, based on our predetermined policy goals.

Other approaches

The foregoing examples are just a few of the methods analysts use to make decisions on complex problems. Many other approaches exist, as an examination of the literature will soon reveal. Or simply develop your own structured approach to decision-making—researchers are continually developing new methods, and yours may contribute to the development of thinking in this area.

5. *How Can I Use My Findings to Reach a Solution?*

This case study is essentially an exercise in structured, objective decision-making. The vast array of issues may confuse and intimidate an inexperienced analyst. The problem-solving structure described in the Introduction can give you a reassuring framework for tackling Varanasi's many issues.

1. *What is the problem?*

In Section 2, we identified the problem as deciding which problems to solve first—sorting out the really critical issues from the less urgent ones.

2. *In what ways do human activities have impact on the natural environment to cause "a problem"? How do these mechanisms give you clues to possible solutions?*

This case describes several (probably too many!) ways in which humans have affected the environment of the Ganges River. We know that some of these activities are commercial and industrial, while others—basic sanitation, drinking water, religious purposes—are deeply rooted in the local culture and religion. Some activities will therefore be more easily curtailed than others. Some, apparently, are causing acute illness and death; others create only a risk of illness in the longer term. This situation, although complex, suggests that some issues may be more urgent than others. (People from most cultures would, for instance, consider high infant mortality rates to be an especially serious concern.) It also suggests that we might want to try easily implemented solutions (such as construction of wastewater treatment plants and clean drinking water sources) before we tackle the more difficult religious issues like cremation.

3. *What governments are responsible for the issue? Whose laws may apply?*

The Ganga Action Plan is a program of the Indian federal government, so that government will certainly have an interest in deciding which problems are most urgent. The local (municipal) government of Varanasi, and the government of the state in which it is located (the state of Uttar Pradesh) will likely also have responsibility here.

4. *Who has a stake in the problem? Who should be involved in making decisions?*

Other than the governments identified in step 3, there are a huge number of possible stakeholders in this case, potentially including local industries, religious groups, commercial interests, residents and tourists, medical professionals, and so on. The list is so long, in fact, that it is really part of the problem: Whose interest is sufficiently urgent to warrant inclusion? Who can safely be omitted, at least for now? It may make sense to work with obvious stakeholders such as the governments to refine a list of issues, then choose additional stakeholders for consultation as problem identification proceeds.

5.   *In the view of your decision-making group, what are the attributes of a satisfactory solution? In other words, when will you be satisfied that the problem is "solved"?*

A satisfactory solution—a satisfactory short list of issues to be resolved—will be one that is widely endorsed by governments and other key stakeholders. It should reflect a societal consensus as to what is urgent and worthy of public expenditure.

6.   *How will you evaluate (test, compare) potential solutions?*

Section 4 presents a number of evaluation techniques that can, singly or in combination, be useful in decision-making. It is essential that these techniques be applied in such a way as to reflect cultural norms and priorities, especially religious attitudes and values, because these are so central to life in Northern India.

7.   *What are all the feasible solutions to the problem?*

Many issues are discussed in this case; there may be many others. Decisions about what is even "a problem" and thus requires intervention should be made with the assistance of a variety of stakeholder and expert opinions.

8.   *Which solutions work "best" in terms of the attributes you identified in (5)?*

Application of the methods discussed in Section 4 should reveal issues that the decision-making group believes are especially important and thus the "best" focus for further action.

9.   *Which solution will be easiest to implement?*

This case study is concerned with setting priorities across a range of environmental issues. If steps 1-8 are applied correctly, a societal consensus should emerge as to which of the observed issues is most pressing and most worthy of public expenditure. Implementation should be smooth for these problems: they are likely to have received extensive media coverage, public debate, and perhaps even preliminary investigation.

10.   *What steps are needed for successful implementation? Who will pay? Who will monitor progress?*

In this case, we are really developing an action plan for a subsequent clean-up program. "Implementation" therefore implies a detailed action plan, with priorities assigned. For example, we might say that the next stage of the Ganga Action Plan should emphasize the provision of clean drinking water supplies and the diversion of remaining untreated human wastes to treatment facilities. We would also assign agency responsibility for these actions and identify funding sources for them. We might set up a plan with actions specified for different time periods: in the next five years, do the following most urgent actions first. Then, in the following ten years, do the next most urgent tasks, and so on. The key here is to translate our decision-making into a series of concrete steps, each of which has timing and financing attached to it—and thus, by implication, priority assigned to it.

6. Where Can I Learn More About the Ecosystem, People, and Culture of the Ganges River?

The following sources contain a variety of information about environmental problems in northeastern India and about the culture and ecology of the Ganges River in particular.

S. Ahmed. 1990. Cleaning the river Ganga: rhetoric and reality. *Ambio* 19(1): 42-44.

K. D. Alley. 1991. On the banks of the Ganga. *Annals of Tourism Research* 19(1): 125-127.

D. S. Bhargava. 1987. Nature and the Ganga. *Environmental Conservation* 14(4): 307-317.

D. S. Bhargava. 1992. Why the Ganga (Ganges) could not be cleaned. *Environmental Conservation* 19(2): 170-172.

A. Gupta. 1988. *Ecology and Development in the Third World*. Routledge, London, U.K.

B. K. Handa. 1992. Water pollution problems in the Indian Subcontinent with special references to the Ganga Action Plan. In: D. I. Coomber, S. S. Langer, and J. M. Pratt (eds.). 1992. *Chemistry and Developing Countries: Proceedings of a Conference, London, April 1991.* Commonwealth Science Council and Royal Society of Chemistry, London, U.K.

United Kingdom

"How can we plan for ongoing protection of natural waters?"

1. What Is the Background?

A need for integrated water management

In the late 1980s, the United Kingdom (England, Scotland, and Wales) undertook an extensive review of its water management framework, including responsibilities for oversight of water resources planning, pollution control, enforcement, navigation, fisheries, and flood protection. This exercise led to a major restructuring of water management responsibilities. Today, the United Kingdom probably leads the world in streamlined, efficient water management.

Under the restructured system, water management planning is the responsibility of the National Rivers Authority (NRA). As part of its mandate, the NRA is now in the process of developing catchment management plans for individual river basins. These plans investigate major water uses (including extractions) and define a community-based plan that ensures that use-related objectives are met.

English water extraction regulations require an abstraction license for any entity taking more than 20,000 liters per day. In the U.K., much of the business of sewage treatment and water supply is

Figure 7.1: Map of the United Kingdom showing location of Upper Wye catchment.

Aberdeen •

• Edinburgh

Belfast •

IRELAND

Dublin •

• Newcastle

• Liverpool

U. K.

Upper Wye catchment

• Cardiff

London •

now "privatized"—that is, operated by private companies under contract to government. The activities of the water companies, who are the primary extractors of potable water, are supervised by the (national) Office of Water Services. The water companies must meet drinking water and pollution standards, as enforced by Her Majesty's Inspectorate of Pollution. The National Rivers Authority (NRA) also has enforcement capabilities for abstractions and outfalls as they influence the river system as a whole. Although this system may seem confusing to the uninitiated, in practice the various agencies have clear and separate mandates that streamline enforcement and improve efficiency.

Water management on a watershed basis

A key feature of the U.K.'s approach to water management is management on a watershed basis. Although it may make intuitive sense to make decisions in this way, much more commonly, water management is conducted on the basis of political divisions—by state or province, for example, rather than by river basin. The reasons for this are obvious: a single major river like the Amazon or the Rhine may pass through several states or even several countries, so a coordinated watershed plan would require that consensus be developed among all jurisdictions—often very difficult, especially where different countries are involved.

Yet there is growing agreement throughout the world that water management on a watershed basis makes the most sense from an environmental perspective. Countries like Canada, the U.S., Mexico, India, and Australia are increasingly trying to incorporate watershed, or subwatershed, management in their water planning activities. In some cases, watershed plans have been expanded into comprehensive regional development schemes, including not only environmental considerations but also land-use planning, road construction, schools, hospitals, industrial development, agriculture, and related activities.

> *Watershed (n.): The line of separation between waters flowing to different rivers or basins or seas; catchment area; ridge of high ground. (The Concise Oxford Dictionary)*

Is watershed management appropriate for all river basins? Probably not. Watershed management may be best applied where the watershed has not only water linkages but also a strong common identity and shared economic, social, political, and cultural interests that coincide approximately or exactly with watershed boundaries. So the Nile or the Amazon, with their immense basins and diverse cultural and jurisdictional bases, might be difficult (although not impossible) subjects for watershed planning.

This general consideration of watershed "homogeneity" has been one of the driving forces behind *sub*-watershed planning in Ontario (Canada), and other countries around the world. The idea is that the basin as a whole may be overwhelmingly complex, hydrodynamically, politically, economically, or culturally, but *parts* of the basin can be managed effectively as discrete units contributing to the whole.

The Upper Wye catchment

The River Wye is one of the United Kingdom's most important rivers, draining the south-central part of Wales and portions of western England. The river passes through varied country, including hilly upland areas (and the old spa town of Llandrindod Wells, Wales), rolling agricultural lands, and industrial towns. Its complex land uses, and the international extent of its basin, make it essential to plan future uses of the basin carefully.

In June, 1993, the National Rivers Authority began a catchment planning exercise for the upper portion of the River Wye basin. The basin has a total area of over 4,000 km^2. For planning purposes, the total area has been divided into two subbasins: the Upper Wye catchment (the River Wye and its lakes and tributaries down to the city of Hay on Wye) and the Lower Wye catchment (the remainder of the basin down to the mouth of the river at its confluence with the River Severn, near the Bristol Channel).

Box 7.1: The Mar del Plata action plan

In 1977, a special United Nations conference on world water resources, held at Mar del Plata, Argentina, resulted in a comprehensive set of recommendations targeted at meeting the goal of safe drinking water and sanitation for all human settlements by 1990. Even then, it was clear that water resources would be increasingly under siege as the need for economic development came into conflict with the desire for protection of the environment.

1. Each country should formulate and keep under review a general statement of policy relating to the use, management, and conservation of water as a framework for planning and implementation. National development plans and policies should specify the main objectives of water-use policy, which in turn should be translated into guidelines, strategies, and programs.

2. Institutional arrangements adopted by each country should ensure that the development and management of water resources take place in the context of national planning, and that there is real coordination among all bodies responsible for the investigation, development, and management of water resources.

3. Each country should examine and keep under review existing legislative and administrative structures concerning water management and should enact where appropriate comprehensive legislation for a coordinated approach to water planning. It may be desirable that provisions concerning water resources management, conservation, and protection against pollution be combined in a unitary legal instrument. Legislation should define the rules of public ownership of water and of large water engineering works, as well as provisions governing land ownership problems and any litigation that may result from it. It should be flexible enough to accommodate future changes in priorities and perspectives.

4. Countries should make necessary efforts to adopt measures for obtaining effective participation in the planning and decision-making process involving users and public authorities. Such participation can constructively influence the choice between alternative plans and policies. If necessary, legislation should provide for such participation as an integral part of the planning, programming, implementation, and evaluation process.

The Mar del Plata action plan, as it is known, emphasized a strong, centralized, national commitment to water management. Yet even 20 years later, the problems it was intended to solve remain significant: the dominance of unregulated water uses; inadequate and ineffective water resource management; inefficiency in water-related public utilities; too few trained staff of all types; overcentralization and bureaucratization of decision-making authority; and inappropriate and inadequate water legislation. Current authors such as Terence Lee believe that centralized water planning has failed as a tool for achieving optimum use of the water resource—just as many believe it has failed as a general tool for achieving social and economic development. Mr. Lee believes that "new emphasis should be placed on the need for achieving rational, efficient use of water locally," and that water management institutions must be "appropriate to local conditions and not centrally, inflexibly, imposed." This perspective, widely endorsed, seems to support the notion of water management on a watershed, not state or national, basis.

Sources: United Nations (1977) and Lee (1992).

2. *What Problem Are We Trying to Solve?*

What is a desirable condition for the catchment?

The purpose of watershed (or catchment) management is to ensure that natural waters such as rivers, lakes, coastal waters, and groundwater are protected and where possible improved for the benefit of future generations. With increasing development, a watershed typically experiences an increasing number of uses, many of which interact or conflict. A good watershed plan balances the needs and objectives of the various members of the watershed community in a way that promotes overall improvement of the water environment. Arriving at this balance of activities—indeed, consensus on a desirable condition for the water environment—is the problem facing the watershed planner.

At first, this seems an overwhelming prospect. In approaching it, the NRA chose to break down the overall problem into several specific questions. These questions can then form the basis for development of an integrated plan that meets the needs of many users. The NRA's key questions were as follows:

1. *What are the catchment uses?*

As a first step in defining the problem—and its possible solution—the NRA elected to identify and describe each use of the water environment within the Upper Wye catchment. Maps are helpful in showing the physical relationships of the various uses.

In addition to describing the use, the NRA recommends developing objectives for each use in terms of desired water quality, necessary water quantity to support the use, and any consideration of physical features relating to the use.

2. *What are appropriate targets for the catchment?*

When all types of users have been identified and described, it may be possible to set overall targets for water quality and water quantity for the catchment. Similarly, targets for physical features may be set, but in this case it may be best to relate targets to individual specific uses.

3. *What is the current "state of the catchment"?*

Once ideal values—the targets—have been identified, the analyst should evaluate the current "state of the catchment" to determine if and where existing conditions are not consistent with desired levels. Again, maps can be helpful in this regard, allowing the analyst to delineate (for example) stream reaches with impaired water quality, or groundwater aquifers with depleted flows. Increasingly, watershed managers are relying on Geographic Information Systems (see Box 7.2) to store and analyze watershed data in this kind of situation.

4. *What specific issues and options exist?*

By this point, the nature of "the problem"—or problems—in the watershed should be beginning to be apparent. Key issues will be emerging, often resulting from conflicts between water users. For example, a given reach of stream may receive the effluent from a sewage treatment plant, but the quality of that effluent is currently creating problems for fish habitat and reproduction.

As specific issues are identified, options for resolving these problems— with their advantages and disadvantages—may also begin to emerge.

It is important to identify the party or parties—whether individual, corporation, or public agency—involved in each issue and/or with responsibility for remedial options.

5.   Who should be involved?

In almost every watershed, a number of different stakeholders are involved directly or indirectly. A watershed plan developed solely by government—or some other party—will therefore not necessarily reflect the interests of all parties adequately. Most effective plans make extensive use of public input in answering the four previous questions.

Sometimes, as in the Upper Wye Catchment Management Plan process, government, in this case the NRA, develops a draft plan that then becomes the basis for consultation between the NRA and all those with interests in the catchment. In the Upper Wye process, consultees were encouraged to comment on the issues and options identified in the plan, suggest alternative options for resolving identified issues, and raise additional issues not identified in the plan.

Following the consultation period, the NRA considered comments for incorporation into a revised final plan that will then form the basis for the NRA's actions within the catchment and for the agency's interaction with other organisations.

The NRA, like other agencies involved in watershed planning, expects that the finished plan will influence not only the agency's own action plans and statements of policy but also those of developers, planning authorities, and others involved in day-to-day management of the catchment.

3. *What Components of the Environment Are Affected, and How?*

Water resources of the Upper Wye catchment

The Upper Wye catchment is primarily rural in character, with the river flowing from an elevation of 741 m above sea level at Pen Pumlumon Arwystli to 71.5 m by Hay-on-Wye. The Upper Wye portion of the river occupies a distance of about 93 km, a little more than one-third of the river's total length of 250 km. The watershed is oriented roughly northwest to southeast, the Upper Wye subbasin making up about 39% of the total catchment area. Local geology is primarily sedimentary rocks (sandstone, mudstone, limestone and shale) of Ordovician and Silurian age. In most areas, bedrock is covered by more recent sands and gravels that are thin and highly permeable.

Rainfall

The Upper Wye catchment is relatively wet compared to many other areas of the world, receiving about 1,350 mm of rainfall per year, of which about 470 mm is lost by evaporation and transpiration. Rainfall is highest in upland areas (up to 2,400 mm/year) and lowest at Hay-on-Wye (about 1,200 mm/year), although as a basin the average annual precipitation is close to the average for all of Wales (1385 mm/year). Taken together, England and Wales average somewhat less precipitation, at 912 mm/year.

Groundwater

Groundwater resources in the Upper Wye basin are important for several reasons. First, they are used as a source of drinking water for many people in rural areas. Second, they act as a reservoir, releasing water to lakes and streams gradually over the year. Sustaining groundwater is therefore important for the preservation of wetlands and bog areas, as well as base flows in rivers and streams.

Bedrock that is highly permeable or fractured tends to carry the most groundwater. Overlying soils are thin, so the water table is often located close to the surface. Sands and gravels overlying bedrock are usually in contact with surface waters, and thus waters flowing through them are vulnerable to the impacts of surface water pollution.

Surface waters

The quality of lake and river waters in the Upper Wye basin is generally very good, probably because of the area's low population density and minimal industrial development. Surface waters are used in some areas for potable (drinking) water and also support a healthy salmon and trout fishery. Some rivers and streams are acidic relative to desirable conditions.

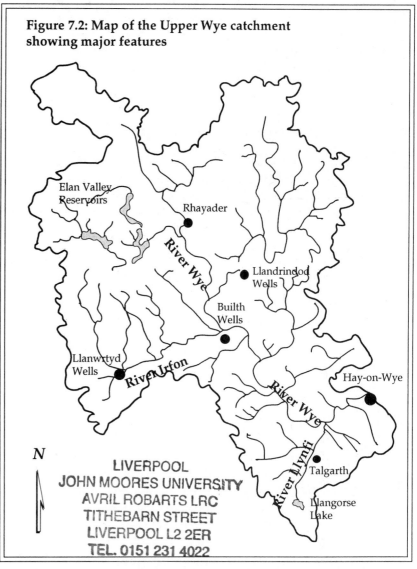

Figure 7.2: Map of the Upper Wye catchment showing major features

Flood defense

As in many other river basins, the flood plain (zone prone to flooding in high flow periods) of the Wye is an appealing place for residential and other development. Although such development has in some cases been permitted in the past, it should probably be discouraged in future. The flood plain is also an important part of the river ecosystem, providing a corridor for movement of animals and a sensitive habitat for a variety of organisms.

Flood defenses (dikes, dams, etc.) protecting existing development have been erected in a number of communities. Typically, these defenses are designed to protect against a flood that would occur only once in 50 years (expressed as the "return period" of the flood). Different land uses may be protected against different sizes of flood, with the most extensive protection in urban areas, where the risk to human life and property is greatest. Agricultural areas may be protected against a more frequent flood. A strong flood-warning system is an essential adjunct to physical defenses against flooding.

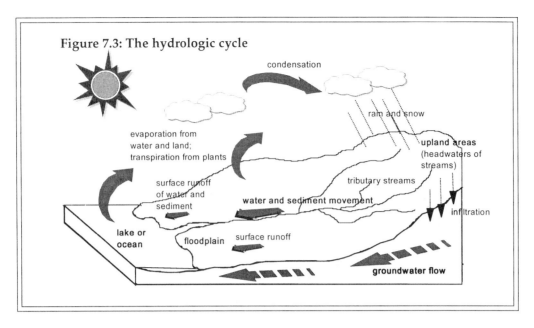

Figure 7.3: The hydrologic cycle

condensation

rain and snow

evaporation from water and land; transpiration from plants

upland areas (headwaters of streams)

tributary streams

surface runoff of water and sediment

water and sediment movement

infiltration

lake or ocean

floodplain

surface runoff

groundwater flow

Land use and infrastructure

The Upper Wye watershed exhibits a variety of land uses. Agriculture is predominant, including sheep and dairy farming, crop production, and moorlands. There is a small amount of mining activity and a larger forestry industry in upland areas. A good network of roads and a major railway line serve the larger centers of Builth Wells (population 2,040), Llandrindod Wells (population 4,943), Llanwrtyd Wells, Rhayader, and Talgarth, and a number of smaller villages. Several of the urban centers are tourist attractions (a result of their historical development as spa towns), so there is a seasonal influx of population during the summer months. Normal population density is about 18 people per square kilometer.

Conservation

The area supports a wide variety of plants, mammals, fish, and birds. Some areas of the catchment have been designated as Environmentally Sensitive Areas, and the River Wye itself is a "Site of Special Scientific Interest" under national law. The area also contains part of a national park.

Water companies

Water supply and sewage treatment in the area is provided by two companies, Welsh Water (Dwr Cymru) and Severn Trent Water. The activities of these companies, including the volume of their water abstractions and the quality of their finished water, are overseen by Her Majesty's Inspectorate of Pollution, as described earlier.

Box 7.4: Length of River Wye and tributaries in National Water Council water quality classes

Class 1A (very good)	380.6 km
Class 1B (good)	68.8 km
Class 2 (fair)	0 km
Class 3 (poor)	4.7 km
Class 4 (bad)	0 km

Class 1A: *Suitable for drinking, game, sensitive fisheries, high amenity value.*

Class 1B: *Quality not as high as Class 1A but usable for substantially the same purposes.*

Class 2: *Suitable for drinking after advanced treatment; supporting reasonably good coarse fisheries; moderate amenity value.*

Class 3: *Polluted to an extent that fish are absent or only sporadically present; may be used for low grade industrial abstraction.*

Class 4: *Grossly polluted and likely to cause nuisance.*

Catchment uses

The NRA divided catchment uses in the Upper Wye into several main classes. These are:

Development and land use

The present Powys Structure (land use) Plan includes provisions for an additional 5,000 residences to be built in the the catchment area by the year 2006. In the towns of Rhayader, Llandrindod Wells, and Builth Wells there may also be a need for localized development of industrial, storage, and warehouse facilities. Although 128 Scheduled Ancient Monuments exist within the catchment, very few of these are affected by river activities. A few, however, are located on flood plains or close to streams.

Forestry

Upland areas of the catchment are forested primarily with coniferous species. In some areas, particularly to the west and north, forest tracts are extensive. Activities such as logging (or even replanting) in upland areas have the potential to impact water quality in the headwaters of the Wye and its tributaries, especially in areas where soils and streams are already acidified and where sensitive salmon fisheries may exist. Tree-felling may also reduce the ability of upland areas to retain moisture, thereby affecting groundwater flows or creating changes in the patterns of flow. Clearly, forestry activities are an important source of impact on wildlife habitat.

Farming

The catchment area contains hundreds of farms, many located on the banks of rivers.

Conservation

A River Corridor Survey undertaken in 1992 mapped the areas of greatest ecological interest in the Upper Wye catchment. The river itself is designated as a national Site of Special Scientific Interest as a largely natural and unpolluted major river. Its varied plant and animal populations reflect the variety of natural habitat types along its length and give it great aesthetic and ecological value. Species of particular interest include certain wetland species (e.g.,wild chives), river otters, the rare Little Ringed Plover, and other valued bird species such as dippers and kingfishers. The advent in recent years of mergansers and goosanders has raised concern about the potential for these fish-eating birds to reduce the populations of young salmon and trout in sensitive areas. The catchment also supports a varied community of insects and aquatic invertebrates typical of pristine water systems. Among these invertebrates are native crayfish, a species for which the area is considered nationally important.

Fisheries

The Upper Wye is a major spawning and nursery area for salmon. According to the NRA, it is arguably the best salmon fishery in England and Wales. During the spawning season, illegal fishing of salmon occurs, with the effect of reducing fish stocks directly, through taking of adult fish, and indirectly, by reducing the spawning population. There is a lively market for salmon from the river, whether caught legally or illegally, and this market may tend to encourage poaching (illegal fishing). In recent years, the salmon catch has declined in the Upper Wye, suggesting that the numbers of salmon in the stream have also declined. If real, this decline may be a result of various factors including illegal fishing, legal fishing, acidification of waters causing depleted phytoplankton and zooplankton stocks, forestry practices, avian predators, physical barriers to migration, and weather conditions. Other important fish species include brown trout and rainbow trout, both of which are stocked, eels, shad, and a variety of "coarse" (as compared to sport) fish including carp, chub, dace, pike, and grayling. Commercial fishing is limited to an eel trap maintained by the NRA on a tributary of the Wye.

Abstractions

Water abstractions are regulated by the NRA under a permit system. Small abstractions, such as residential wells, do not require a permit. The area's groundwater resources are an important resource for Dwr Cymru, the local water company. Abstracted water is mostly returned to the catchment via discharges of sewage treatment plant effluent to rivers and streams. Surface water is also abstracted for drinking water supply, primarily from reservoirs, some of which are used for local electricity generation (only possible when the reservoirs are full). Some increase in potable water demand is expected as the area develops further, but this increase is expected to be small. Reservoir water is also released to the river during low flow periods under an agreement between Dwr Cymru and the NRA. Extensive "draw-down" in the reservoirs has been a matter of concern to ecologists because of its potential to affect aquatic biota.

Agricultural abstractions occur throughout the watershed for general agricultural use (including livestock watering), spray irrigation, and fish farms. About 60% of general use water is returned to the river after use, almost all of the fish-farm water (which can be large in volume), but virtually none of the spray-irrigated water.

Industrial water use is limited in the Upper Wye, because of the area's largely rural character. Two abstraction licenses have been granted for sand and gravel washing operations and related activities. A third license allows groundwater abstraction for industrial use. Future growth in demand is hard to predict but will likely be less than 1% per year.

Discharges and pollution control

There are 39 sewage treatment works operating in the Upper Wye catchment, most of them owned by Dwr Cymru. All are monitored by the NRA. In urban areas, there are several stormwater overflows (drainage from streets and roofs) but the impact of these is thought to be negligible. There are very few industrial effluent dischargers in the basin. Filter backwash is discharged from Dwr Cymru's water treatment plants, and there is a small amount of discharge of effluent from fish farms, quarries, and sawmills. Spillage of chemicals including chlorine (used in water and wastewater disinfection) is always a possibility.

Three landfill sites are located in the basin, but none is located close to a stream and so they are not thought to pose any threat to water resources.

Amenity, navigation and water sports

The appearance of the river is important to residents and others who visit the area. Recreational activities in the basin include walking, boating (primarily white-water canoeing), sailing and water skiing on the lakes, and a very limited amount of swimming (discouraged by the NRA).

Box 7.5: What about the future?

The NRA's catchment management plan for the Upper Wye centers on current problems and modest projects of growth over the next decade. What actions should they take to assure longer-term preservation of the Upper Wye's water resources?

4. How Can I Analyze This Information?

Matching uses with existing quality

The main analytical challenge in this problem is to match uses with existing quality. In this way, it should be possible to identify key issues and options for resolving them. These issues can be identified in many ways, but it should be emphasized that the analysis is a qualitative one, typically based on "brainstorming", rather than a quantitative one.

In the case of the Upper Wye, the Welsh Region of the NRA developed a preliminary list of issues and options that it then circulated throughout the community for comment. In some other processes, such as that in the Grand River, Ontario, and in the Remedial Action Plan process (a joint Canada-United States initiative to manage pollution "hot spots" in the Great Lakes), the public is directly involved in the definition of issues and options.

The NRA's issues, and options for their resolution, were as follows:

1. *Acidification.* This results from the deposition of acid precipitation (caused by the burning of fossil fuels) on acid-sensitive soils and "soft" waters (waters low in calcium carbonate, a natural acid-neutralizer or "buffering" agent). Acidification causes problems for the survival of fish eggs and young fish, reduces aquatic species diversity and thus may affect carnivorous species like otters and fish-eating birds, and may increase the need for treatment of drinking water. Options may include restricting emissions of acid-causing gases from smokestacks, choosing species other than conifers for reforestation, and possibly adding lime (calcium carbonate) to raise pH in natural waters.

2. *Impaired fisheries resulting from low dissolved oxygen in some areas.* The sources of—and thus the solutions for—this issue are not presently clear. The problem may be a result of pollution from industrial and/or agricultural activities.

3. *Blue-green algae in lakes and ponds.* Algae blooms create water-quality problems by using large quantities of oxgyen at night, through respiration, and through the oxygen-demanding decay of dead plant tissue. Here again, there is little information available on the causes of the problem, although they are likely related to enrichment of natural waters with nutrients, especially phosphorus and nitrogen. These materials may enter water through a variety of human activities including sewage discharge, agricultural runoff, and industrial effluents.

4. *Low flows in summer months.* Portions of the river exhibit very low flows during the summer, the time when the demand for spray irrigation is greatest. Water used in spray irrigation is not returned to the river, so this type of abstraction can have a significant negative effect on river flows. Solutions include storage of more water in reservoirs through the winter months so that more is available for release in summer. Farmers can also construct on-farm reservoirs, or grow crops that are less dependent on spray irrigation.

5. *Environmental impacts of abstraction are not known.* It makes sense that abstractions of water should be stopped when they begin to affect the river environment. However, current understanding of these impacts, and when they occur, is very weak. Additional research in this area is needed.

6. *Impacts of development of base flows are not known.* Some people believe that land drainage and land-use changes have altered the natural flow regime in the river, making it "flashier"—faster to respond to rainfall—and lower in its base flows than in the natural condition. Again, more research into these possible impacts is needed to clarify the extent of hydrologic changes following development.

7. *Need to protect and enhance the wildlife resource.* Human activities affect wildlife habitat in many ways. Protective measures could include consideration of wildlife requirements when approving abstraction licenses, land drainage permits, discharge permits, and land-use-planning applications. River-edge vegetation should be protected where sheep and cattle trample banks where they enter the river to drink. The impact of recreational activities on wildlife should be evaluated carefully, with a view to limiting certain activities when and where necessary.

8. *Decline in salmon and brown trout stocks.* Over the past 20 years, the NRA has observed a decline in the number of spring salmon and brown trout caught on the Wye. The reasons for this decline are not clear but are likely to be complex, including acidification, habitat degradation, physical barriers to migration, and overfishing. Controls or improvements in all these areas are possible and should result in increased fish stocks. The impact of avian predators can be evaluated through separate research.

9. *Flooding at Builth Wells and Llanelwedd.* Flooding of a main road through these communities occurs in floods with a return period of greater than 1 in 4 years. Thirteen homes and 20 commercial properties are affected by this flooding. Flood defenses have been considered but were rejected as too costly and environmentally undesirable.

Identify areas of conflict

The above list illustrates some areas where existing uses match or conflict with existing quality in the Upper Wye catchment. It also gives some indication of where problems are most urgent, where additional information is required, and where concerns are probably minor. It does not give a good picture of the future: changing land uses and pressures on catchment water resources. This would be an important part of any comprehensive analysis.

Project the impact of management actions

Effective planning for the Upper Wye catchment should include projections of future uses, their intensity, and their impacts on catchment water resources. It should also examine the effect of possible management actions that could be undertaken—actions like curtailing certain uses or use intensities. This kind of prediction is best made using a computer simulation model of the type described in Case Studies 5 and 11. Such a model can simulate streamflow and water quality throughout the basin under various existing and future scenarios, and thus guide the analyst as to which actions will be most effective in meeting catchment targets.

5. How Can I Use My Findings to Reach a Solution?

Use the decision-making framework described in the Introduction to organize your thinking on this problem, as follows:

1. *What is the problem?*

In Section 2, we identified the problem as finding consensus on a desirable condition for the water environment and a sustainable balance of uses in the catchment area.

2. *In what ways do human activities have impact on the natural environment to cause "a problem"? How do these mechanisms give you clues to possible solutions?*

The Upper Wye catchment supports many different uses and users which affect the watershed environment in a variety of ways. The detailed list of uses given earlier gives some indication of the kinds of impacts currently observed. As in Case Study 4, it appears that some uses may be incompatible with others. In other cases, the intensity of use may be too great. This guides us to a determination of incompatible uses, and decisions about which will be allowed to continue. It also suggests that, where uses are allowed to continue, the intensity (or perhaps the timing, or both) of use may need to be restricted.

3. *What governments are responsible for the issue? Whose laws may apply?*

Recent legislative reform in the U.K. has greatly clarified adminstrative responsibilities. The National Rivers Authority now has clear authority over catchment planning, although some other agencies (as discussed in earlier sections) may have responsibility in certain well-defined areas.

4. *Who has a stake in the problem? Who should be involved in making decisions?*

Catchment stakeholders and potential decision makers could include any resident of the catchment, including private landowners, industrial operations, municipal operations such as sewage treatment plants, transient recreational users and tourists. Regulatory agencies, particularly the NRA, will also have a stake in the problem.

5. *In the view of your decision-making group, what are the attributes of a satisfactory solution? In other words, when will you be satisfied that the problem is "solved"?*

The NRA clearly state that target-setting is a catchment responsibility. Ideally, the entire river should be water-quality Class 1A, so this is probably the ultimate goal of the plan. Uses should be sustainable—that is, should be able to continue indefinitely without negatively affecting the quality or quantity of catchment waters. We could therefore set specific targets for water quality (for example, target concentrations for individual pollutants) or flows, and track progress against those targets.

6. How will you evaluate (test, compare) potential solutions?

Computer simulation models could be a useful tool in predicting the impact of management actions on water quality and quantity. These models are discussed in more detail in Case Studies 5 and 11. Consensus-building approaches such as conflict resolution (see Case Study 4) could also be very useful in reaching community agreement about targets, desirable uses, and remedial measures. Evaluation could include both quantitative techniques such as computer simulation and cost-benefit analysis, and qualitative techniques such as environmental impact assessment (see, for example, Case Studies 15 and 16).

7. What are all the feasible solutions to the problem?

There are innumerable combinations of uses and use intensities for the Upper Wye. Developing a list of feasible approaches is a matter of creativity and consultation with knowledgeable individuals from this or similar watersheds. Examination of the literature can reveal parallel cases where, for example, certain combinations of uses have been unsuccessful or where innovative methods have been used to reduce the impact of existing uses.

8. Which solutions work "best" in terms of the attributes you identified in (5)?

The "best" solution will be one that is acceptable to the community and achieves long-term protection or improvement of catchment water resources. There are many solutions that could provide long-term protection, but few will be readily accepted to the community or sustainable over a long period of time. Solutions that meet these requirements will probably therefore emerge as "best".

9. Which solution will be easiest to implement?

Implementation obstacles can only be identified when a concrete proposal has been developed. It is likely that proposals that eliminate certain valued activities, for instance recreational fishing, would meet with considerable social protest. Similarly, proposals that are costly—especially those which impose the majority of cost on one user group—will be less acceptable than those that distribute the cost burden among many users.

10. What steps are needed for successful implementation? Who will pay? Who will monitor progress?

The variety of uses in the basin, and the need for long-term protection, mean that community acceptance is critical to successful implementation. The details of implementation will depend on the individual proposal (as in Case Study 4, these could be developed through role play exercises). What may be more important in this case is the process used to make decisions and ultimately to secure community and agency agreement to the plan.

6. *Where Can I Learn More About the Ecosystem, People, and Culture of the River Wye Basin?*

The following sources give useful insight into the problem of watershed management and the particular environment of the Upper Wye catchment.

A. S. Goodman and K. A. Edwards. 1992. Integrated water resources planning. *Natural Resources Forum* 16(1): 65-70.

T. Lee. 1992. Water management since the adoption of the Mar del Plata Action Plan: Lessons for the 1990s. *National Resources Forum* 16(3): 202-211.

N. I. McClelland. 1987. Improved efficiency in the management of water quality. *Natural Resources Forum* 11(1): 49-57.

Adrian T. McDonald and David Kay. 1988. *Water Resources Issues and Strategies*. Longman Scientific and Technical, London, U.K.

National Rivers Authority. 1993. National Rivers Authority Strategy (8-part series encompassing water quality, water resources, flood defense, fisheries, conservation, recreation, navigation, research and development). National Rivers Authority Corporate Planning Branch. Bristol, U.K.

National Rivers Authority. 1993. Upper Wye Catchment Management Plan Consultation Report. National Rivers Authority, Welsh Region. Cardiff, Wales.

G. Schramm. 1980. Integrated river basin planning in a holistic universe. *Natural Resources Journal* 20(1980): 787-806.

United Nations. 1977. Report of the United Nations Water Conference, Mar del Plata, 14-25 March 1977. E/CONF 70/79, United Nations, New York.

W. Viessman, Jr. 1990. Water management issues for the nineties. *Water Resources Bulletin* 26(6): 883-891.

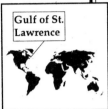

"How should we estimate the size of the harp seal population so we can make good management decisions?"

1. What is the Background?

A valued resource from ancient times

One of Atlantic Canada's commonest seal species, *Phoca groenlandica* (the "Greenland seal"), is best known as the harp seal, in reference to the dark horseshoe shape on the back of adult males and most females. Other females are gray with smaller dark patches whereas the infants are covered with fluffy white fur for the first 2 weeks of life.

The distribution of these mammals ranges from the White Sea, in the Arctic Ocean, around Greenland, and west to the Gulf of St. Lawrence and northern Newfoundland. The majority of the population is in the Newfoundland stock, but total population size is currently unknown.

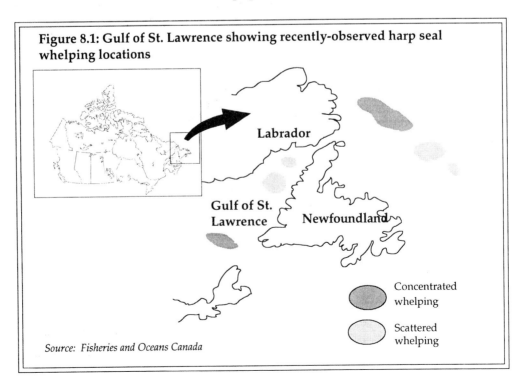

Figure 8.1: Gulf of St. Lawrence showing recently-observed harp seal whelping locations

Labrador

Gulf of St. Lawrence

Newfoundland

Concentrated whelping

Scattered whelping

Source: Fisheries and Oceans Canada

1661 *French merchant Francois Bissoit obtains sealing rights for the St. Lawrence River.*

1765 *Jeremiah Coughlan establishes the first English sealing post in Labrador. Within 5 years, pelts are worth 10,000 pounds sterling a year.*

1799 *The seal hunt now takes more than 120,000 animals in St. John's and Conception Bay, Newfoundland.*

1828 *The first Halifax-based vessels begin hunting seals in the Gulf of St. Lawrence.*

1844 *Seal products account for one-third of the value of Newfoundland exports. About 18 million seals are killed between 1800 and 1860.*

1858 *There are now 90 Nova Scotian boats killing over 30,000 seals a year in the Gulf of St. Lawrence.*

1881 *Concerns are raised about the declining numbers of seals.*

1915 *Canadian Fishermen magazine predicts the extermination of seals by mid-century unless they are protected.*

1939 *Norwegian vessels join the seal hunt in the Gulf of St. Lawrence.*

1949 *A survey estimates the number of harp seals at 3.3 million.*

1960 *A survey finds the number of seals has dropped to 1.25 million*

1964 *Tough new regulations limit the number of pups taken to 50,000, require licensing of hunters, and prohibit the skinning of seals alive.*

1977 *Canada introduces a quota allowing 160,000 seals to be killed, 152,000 of them pups.*

1983 *Europe, formerly 75% of the export market for seal products, bans the importation of products made from Canadian seal pups.*

1988 *The Canadian government bans the killing of baby seals (less than 1 month old) but allows a land-based hunt for adults.*

1990 *A survey reports about 3 million seals in the North Atlantic. About 50,000 adult seals are killed annually.*

1994 *Markets for seal products have declined significantly. Newfoundland lands a deal to sell 50,000 seal carcasses to China. Concerns are raised that the animals are being killed solely for their penises, for use in aphrodisiacs.*

*Adapted from Kevin Cox, "Centuries of slaughter," Toronto **Globe and Mail**, March 12, 1994.*

Northern Canadian native peoples have been hunting seals for their meat and fur for centuries. The oils from seal blubber are used in lighting, heating, cooking, and lubrication. Flippers are sold as food in Newfoundland. Carcases are often sold for use in animal feed. Probably the most controversial use of seals has been for their fur, which is made into various articles. Also controversial is the sale of seal penises to certain Asian countries for use in aphrodisiacs.

One of the most difficult aspects of the seal hunt is the problem of defining a sustainable hunt quota. One approach would be for the federal government to establish a quota that allows a sustainable harvest of this renewable resource. This quota has been difficult to establish and has been revised many times over the history of the seal hunt.

Although sealing has been taking place in Atlantic Canada since as early as 1661, the hunt remained unregulated by the Canadian government until the early 1960s. In 1960, a survey found that the seal population in the Gulf of St. Lawrence had declined to about 1.25 million animals from the 3.3 million recorded in 1949. As a result, in 1961, the Canadian government introduced regulations requiring that sealers be licensed, that they not skin seals alive (until that time a common practice), and that the total hunt of seal pups in the Gulf of St. Lawrence be limited to 50,000 animals. (In 1987 the large-vessel commercial hunt of "white coats"—infants—was banned.) By 1977, Canada had introduced a quota allowing the harvest of 160,000 seals; by 1996, the Atlantic Seal Management Plan increased the seal harvest quota to 250,000 animals.

 # 2. *What Problem Are We Trying to Solve?*

Many issues

There are many issues—social, environmental, and economic—surrounding the exploitation of harp seals for economic use. Animal welfare groups express concern over the seal hunt, both for the well-being of the population and for humane treatment of the animals.

The environmental impact of the seal hunt is also an important issue. Many researchers and organizations have attempted to assess these impacts and have found it a challenging task. In part, this is because of the enormous complexity of the marine food web in which the seals play an important role.

Marine scientist David Lavigne has said...

"In fisheries management, as in wildlife management, there is a tendency to look to 'science' for all the answers—even if, down the road, the scientific advice is often neglected or conveniently forgotten. In any case, the fact is, 'science' does not have all the answers, nor will it ever. And for that reason, we might do well to consider other sources of information. In the case of overfishing, there is, for example a rich history from which we might just learn some useful lessons."

Source: *BBC Wildlife*, May 1992

The decline of the fishery

In recent years there has been a dramatic decline in the in the populations of groundfish such as cod, and other fish species such as capelin, on the Atlantic coast of Canada. Historically, the groundfish industry has been an economic mainstay of eastern Canada. Its demise therefore has serious social and economic consequences for that area, and its recovery is eagerly anticipated by the industry and the government alike. Perhaps because of this urgency, there is considerable pressure on scientists and regulators to identify the causes of the collapse. Because the seals' diet is predominantly fish (although not necessarily commercial fish species), some Canadian government officials believe that seals inhibit the recovery of the fishery. This reasoning underlies recent increases in the seal hunt quota, from 186,000 to 250,000.

This increase in the hunt has been highly controversial, not only among animal welfare activists but even among the hunters themselves. The current world market for seals is very limited, and sealers fear that increasing the quota can only put further pressure on the seal industry. (The argument has, however, been attractive to those employed in the fishing industry, who view seals as competitors for a scarce resource, and who have already suffered from reductions in their own harvest of fish.)

A number of scientists have argued (based on analysis of seal stomach contents) that Atlantic cod comprise only a small part of the harp seal's diet. Evidence from the Barents Sea suggests that populations of capelin in that area may be controlled more by physical factors such as salinity, water temperature, or wind direction than by predation. It is also possible that certain species may also be experiencing an increase in natural mortality. An even stronger possibility, in the minds of many scientists, lies in overfishing by Canadian and international vessels over the past 100 years. Other scientists argue that if harp seals are not consuming cod, they may be competing with cod for the same prey species, especially capelin, and thus indirectly having an impact on the cod population by starving them of prey. On the other hand, some research shows that a decrease in the harp seal population could mean a decrease in the populations of other species such as squid, a known prey item of seals. Fewer seals could therefore mean more squid available to prey on recovering cod stocks.

How many seals?

Scientists also debate the current size of the seal population and its potential rate of growth. Recent population size estimates range from 2.6 million to 4.8 million, with all estimates supported by extensive scientific research. If true, the larger number seems to point to a population growth rate of about 5% in recent years, suggesting that more seals could be hunted without endangering the overall population. Other authors suggest that the higher population estimate is an artifact of estimation techniques, and that "observed" increases simply reflect a redistribution of the population in response to the abnormally low water temperatures and salinity levels observed in recent years.

It is clear from this debate that the causes of the fishery decline, and the role of seals in that decline, are probably diverse. Indeed, some scientists have suggested that our understanding of complex marine food webs is so limited that it is unrealistic to expect to "manage" them effectively with tools like hunting quotas.

In this case, we will deal with a very specific problem, that of estimating the size of the harp seal population and the impact of this estimate on environmental decision-making.

3. What Components of the Environment Are Affected, and How?

The biology of the harp seal

Harp seals are a migratory species common in areas of the North Atlantic and are probably the most abundant seal in Atlantic Canada. Their life expectancy is estimated at about 15 to 20 years, throughout which time they follow a predictable annual cycle of migration, pupping, nursing, and mating (see Box 8.2).

Observations of the seals' diet and reproductive behavior are difficult to obtain, simply because the seals spend much of their life in water and out of sight. Pups are nursed on the ice surface, however, so this time in the life cycle provides an opportunity to collect information about pregnancy rates, age distribution in the population, pup production, and levels of disease and infirmity in the population. Available information is collected from a variety of sources, including aerial surveys, visual surveys, catch data, and mark-recapture studies. The range of techniques and observers (scientists, sealers, fishermen, etc.) understandably leads to a range of observations, so there is considerable debate about the accuracy of any reported findings. Nevertheless, the following can be taken as a general guide to harp seal population dynamics.

Reproductive behavior

Adult female harp seals become sexually mature at about 5 years of age. There is some evidence that females are now maturing later than they did a decade or more ago, averaging 5.5 years to maturity in the early 1990s as compared to 4.6 years in the early 1980s and 6.2 years in 1952. There is some evidence that female sexual maturity is density-dependent, acting as a sort of self-regulating mechanism for population size. When populations are large, as they were in the early 1950s, maturity is delayed. When density begins to drop, as it did between 1950 and 1970, maturity occurs earlier.

About 70% of mature females each year are now observed to be pregnant; in the late 1970s this number was closer to 90%. It must be emphasized that these conclusions are based on a very small number of studies and direct observations; *yet pregnancy rates and age at maturity are central factors in estimating seal population size and growth rate.* Some of the observed changes may be due to the movement of females among whelping areas or other behavioral factors.

For whatever reason, pup production is thought to have decreased from 1955 to 1975 but has been increasing steadily since that

---

Box 8.2: A year in the life of a harp seal

November-February	Feeding,energy storage
February	Females give birth (pup) on pack ice
February	Nursing, care of infants
March	Mating begins
March-April	Migration to subarctic
April-August	In subarctic
September-October	Migration to Newfoundland and Gulf of St. Lawrence

---

time. The increase has occurred despite later maturity and lower pregnancy rates, and could be a result of larger population size and a higher total number of breeding females.

Breeding females exhibit a process called suspended embryotic development, in which the egg is fertilized but development is delayed for two to three months (presumably until the migration to the high Arctic, with its extreme energy demands, has been completed).

Feeding behavior

The diet of the harp seal is quite varied, including a wide range of invertebrate species such as shrimps, crabs, sea urchins, molluscs (e.g., squid, octopus, mussels) and a number of fish species such as skate, herring, smelt, cod, sculpin, capelin, and many others. In fact, stomach content analysis shows that harp seals consume over 110 different species of fish and invertebrates.

Analysis of stomach contents also reveals that the seals prefer prey less than 35 cm in length (in other words, fish too small to be of immediate commercial value) and therefore are unlikely to compete directly with commercial fishing activities. Many other species also prey on these young fish, including whales, sea birds, and predatory fish.

It can be argued that consumption of young fish reduces the number of fish available to grow to maturity, but natural mortality of juvenile fish is very high in any case. Some authors argue that seals therefore eat fish that are of low reproductive value and that have a high potential mortality. By contrast, they argue, commercial fishing harvests mature fish of high reproductive value that are therefore most important in terms of recovering fish stocks. Other scientists argue that the seals starve the dwindling cod stocks by competing with them for prey.

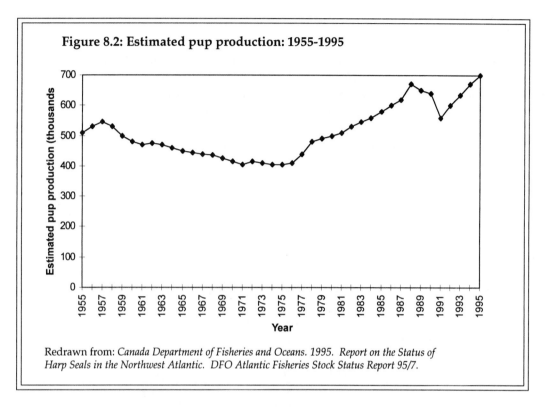

Figure 8.2: Estimated pup production: 1955-1995

Redrawn from: *Canada Department of Fisheries and Oceans. 1995. Report on the Status of Harp Seals in the Northwest Atlantic. DFO Atlantic Fisheries Stock Status Report 95/7.*

Predators of the harp seal

Although the annual seal hunt is a major predation mechanism within the seal population, it is not the only one. Other major predators of harp seals include sharks and killer whales (indeed, some scientists believe that recently observed increases in the population of gray seals can be attributed to increased catches of sharks off Eastern Canada). Predation from sharks may decline in the future if (as expected) interest increases in Atlantic sharks as commercial species.

What is a "sustainable yield"?

The question of a "sustainable yield" has plagued regulatory agencies, hunters, and animal welfarists for decades. This question is more complex than it appears. First, it is necessary to establish the current population size and to be able to compare that population with past observations. If population growth can be established, it may be possible to hunt a certain number of animals. In the present case, the Canadian Department of Fisheries and Oceans (DFO) believes that the harp seal population has increased at a rate of about 5% per year since 1990. They have therefore established a "total allowable catch" (TAC) for 1996 at 250,000, roughly 5% of the population they believe exists. (Other observers challenge DFO's population estimates, and therefore their seal hunt quota.) DFO arrived at this figure using a variety of estimation techniques (see next section) and assumptions about pregnancy

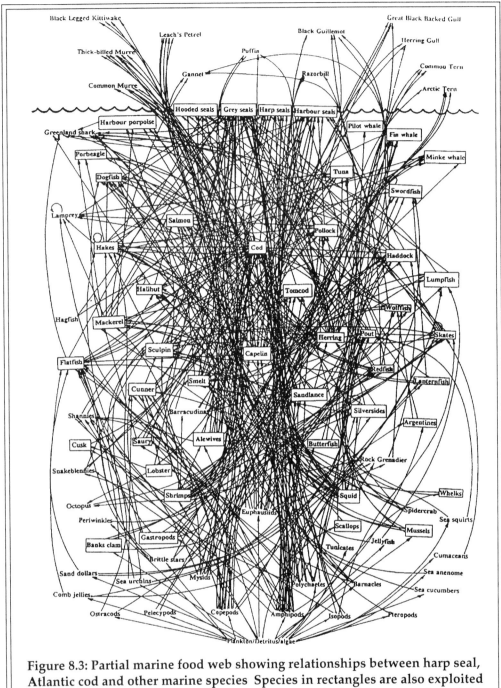

Figure 8.3: Partial marine food web showing relationships between harp seal, Atlantic cod and other marine species Species in rectangles are also exploited by humans. Figure compiled by D. Hucyk. Published with permission of the International Marine Mammal Association.

rates, age of catches, and similar factors. As discussed above, there is considerable uncertainty with these estimates and therefore considerable controversy around their application.

A "sustainable yield" implies that a specific number of animals can be removed from the population year after year, without adverse effects to either the target population or the marine food web. In practice, the concept is tricky because the sustainability of the hunt depends not only on the number of animals taken but also on their sex, age, and fertility. Scientists have been observing subtle changes in the age distribution of the seal population, the age at maturity, and rates of pregnancy. All of these factors may be related to environmental influences or to hunting pressure, and it is difficult to distinguish between the different forces.

But what point is a sustainable yield without a sustainable market for the products of the hunt? Over the past decade, European markets for seal products, formerly 75% of the export market, have all but disappeared as a result of protest against hunting practices, particularly reflected in a European Union ban on trade in seal products. Newer markets, such as those in Asia, may be ethically unacceptable if they use only certain parts of the seal—for instance, the penis—and discard the remainder. Should new markets for seal products be developed? Where, and by whom? At present, increasing the seal quota may flood an already glutted market with unwanted carcasses, reducing profits to hunters without a guarantee of improvement in fisheries or of protection of the seal population itself.

Box 8.3: Difficulties with describing diet

The feeding habits of marine mammals are traditionally assessed by using stomach content analysis and analysis of fecal material. These methods can produce varying results, depending on the time and place of sampling and the rate at which different foods are digested.

It is also difficult to obtain a truly random sample, especially when the sampler must rely on collections from animal carcasses and not on a carefully controlled selection of animals.

Results can also differ substantially depending on whether the analyst uses prevalence (percent of stomachs containing the food item), numbers (of a given food item, expressed as a percentage of the total number), wet mass (percent of total mass) or energy (percent of total stomach contents) as the measure of choice.

 # 4. *How Can I Analyze This Information?*

Estimating population size: an inaccurate science

Population estimates are derived with a variety of techniques. It is this variation in technique, in fact, that creates controversy in estimating population size. Each method has advantages and disadvantages, and new approaches are always under development. Estimates derived from different techniques are very difficult to compare, so it is also difficult to assess population trends over time, when "population" may have been estimated in several different ways.

Despite these estimation complexities, population trends and projections play a key role in decision-making about quotas affecting environmental and economical issues.

The most accurate method of estimating population size is simply to count all the animals: for the harp seal population, this is an impossible task given the wide geographic distribution and large size of the population, and the large proportion of time the animals spend in the water. Various techniques are therefore used to make these estimates. The following are several of the most commonly used population estimation techniques.

Mark-recapture experiments

Mark-recapture experiments are an older method of estimating population size. They have been used on many different types of animals, from rodents to marine mammals.

Their value lies in their ability to produce estimates of population birth and death rates (immigration/emigration rates) as well as population density.

In essence, the technique involves trapping of a given number of animals, the number being based on statistical estimates of population variability. The captured animals are then marked and released. After some time has passed, a second round of trapping is conducted. Several rounds of recapture may occur.

The number of marked, recaptured animals allows us to estimate population density in this way: if we draw a random sample of animals from a marked population, the sample will contain some marked and some unmarked individuals. We can estimate the size of the population by the number of marked animals actually caught (a simple count) and by the proportion of marked animals actually caught (usually estimated with model estimation).

This method is based on three critical assumptions: that marked and unmarked animals are captured randomly, that marked animals experience the same mortality as unmarked animals, and that marks are not lost or overlooked. All these assumptions have created problems in field surveys. Some recapture methods may rely on the efforts of nonscientists such as commercial sealers or fishers, who cannot be relied upon to report findings as carefully as a research scientist might.

Survival indices

The survival index (SI) method provides an estimate of pup production over a period of years for which the pup production rate is assumed to be constant. This method is based on the assumption that an age sample is distributed as a Poisson function (a particular statistical function representing the probability of an event occurring in each of a number of successive time periods). If many years of data are available, the Poisson distribution can be estimated accurately. If only a few years, or 1 year, are available, the estimate is much less reliable. The SI method has come under criticism in the scientific community primarily because of its assumption of unchanging pup production (birth) rates—not a valid assumption for most natural populations.

Quadrat sampling

In the quadrat sampling method, the geographic area of interest—possibly the entire range of a population—is divided into quadrats of known and equal size. All of the animals in a number of quadrats are then counted and the total extrapolated to the total area under study. For this method to be accurate, the sampler must be able to count all the animals in a given quadrat (i.e., not miss any), the area of each quadrat must be known accurately, and the quadrats that are sampled must be representative of the area as a whole (in other words, not likely to support fewer or more animals than the average). Quadrat sampling is widespread in population biology, but it can be difficult to apply to the seal population because of the high proportion of time the seals spend underwater.

Visual and aerial-photographic surveys

Aerial photography has been an important adjunct to seal population size estimation. During the whelping season, the pups remain on the ice to be nursed by their mothers. This is a good time to conduct aerial surveillance using fixed-wing aircraft and helicopters over known whelping areas. The aircraft fly pre-determined transects across known whelping areas, usually in mid-March. Usually a number of transects are flown in the same survey, perhaps 10 to 20 km apart (depending on the circumstances of the survey) and at an altitude of about 250 m above the ice pack. Some surveys fly transects that are intended to overlap, providing comprehensive photographic coverage of a specific area. The length and spacing of transects is determined by statistical techniques to assure a random sample.

For transects where photographs do not overlap, the analyst must make a number of assumptions in analyzing results. Most important of these is that the number of pups in the unobserved portions be the same as that in the observed portions.

Error can creep into these estimates in several ways. To overcome this problem, several analysts sometimes examine the same data set (set of photographs, for example), or reread photographs on several occasions, to reduce perceptual errors. Among the errors that can occur are misidentification of pups (for example, a rock or piece of ice is identified as a pup), and inaccurate estimation of pup age based on pelt characteristics and other physical features. Since surveys are sometimes flown over a period of several weeks, pups can change characteristics during the survey period and be counted in several categories in the same survey period. The apparent age distribution of the population will therefore be different every time a transect is flown.

How should we handle these uncertainties?

The increased seal quota proposed by DFO for 1996 is based on their estimate of an increase in population size between 1990 and 1994. Some scientists, notably David Lavigne, have taken issue with this position on the basis that the two surveys used different sampling designs and analytical techniques. DFO has estimated population size and population growth using a computer model based on results from several years and several different analytical methods, and which assumes that mortality rates are the same for seals of all ages, including pups.

Despite these fundamental disagreements, population estimation is a central part of government policy regarding the seal hunt. Until a way is found to reconcile policy requirements with the error inherent in estimating population size, the debate is bound to continue.

5. *How Can I Use My Findings to Reach a Solution?*

This case presents a problem—estimation based on sampling results—that is common to many cases in environmental studies and ecology. The problem-solving framework described in the Introduction can help you to sort out the issues involved in reaching a decision.

1. *What is the problem?*

In Section 2, we identified the problem as estimating the size of the harp seal population and determining the impact of this estimate on environmental decision-making.

2. *In what ways do human activities have impact on the natural environment to cause "a problem"? How do these mechanisms give you clues to possible solutions?*

The harp seal is a native species, commonly found in northern ocean waters throughout Canada, Russia, and Asia. Human impacts on this population primarily involve hunting but could also include fishing, which may indirectly impact the seal population by depriving it of prey species. This understanding gives us some insight into the forces that affect the size of the seal population and thus areas where the seals may be especially reduced in number.

3.   *What governments are responsible for the issue? Whose laws may apply?*

The Canadian federal government, through the Department of Fisheries and Oceans, is the government directly responsible for this issue. Municipal and provincial governments would likely have little or no role in governing the seal hunt or setting hunt quotas.

4.   *Who has a stake in the problem? Who should be involved in making decisions?*

Possible stakeholders clearly include hunters and processors of the seals, consumers of seal products, and intermediaries such as retailers. The Canadian federal government is also an important stakeholder here. Animal welfare groups and environmental non-government organizations (especially Greenpeace, who first drew public attention to the seal hunt in the 1960s) would also have a strong interest. Finally, the long debate about the seal hunt has prompted considerable scientific interest in the biology and ecology of the seals; marine scientists would also be key stakeholders.

5.   *In the view of your decision-making group, what are the attributes of a satisfactory solution? In other words, when will you be satisfied that the problem is "solved"?*

For the purposes of this case study, a satisfactory solution is a satisfactory estimate. This probably means an accurate estimate, or at least an estimate whose accuracy is known.

6.   *How will you evaluate (test, compare) potential solutions?*

A variety of population estimation techniques are described in Section 4. Each of these has positive and negative aspects. Some may be time-consuming and costly to implement, although they provide a high level of accuracy. Others may require staff with specialized training. You should choose a "best" technique using the approaches described in Case Study 6: set constraints and criteria for your choice, weight criteria if appropriate, then choose the "best" technique.

7.   *What are all the feasible solutions to the problem?*

There is a large literature on population estimation field techniques. Section 6 provides an entry into this literature, while Section 4 describes some of the more common approaches. You may choose from these or develop a new estimation technique of your own devising.

8.   *Which solutions work "best" in terms of the attributes you identified in (5)?*

The "best" choice will likely be one that is cost-effective (provides a good return of information for each dollar spent) and that yields a reasonably accurate estimate.

9.   *Which solution will be easiest to implement?*

Population sampling of the kind discussed in this case is usually the role of regulatory agency staff (in this case, from the Department of Fisheries and Oceans) and/or marine scientists. Ease of implementation should be judged by these individuals, against concerns such as available human and financial resources, personal safety and access, level of training required, and so on. Implementation problems could arise if, for instance, DFO lacked specialized staff required for a certain technique like air photo intepretation. They would then have three choices: use another technique, hire new staff, or train existing staff. This decision would in turn be based on available funding for staff or training, and so on.

10. *What steps are needed for successful implementation? Who will pay? Who will monitor progress?*

Designing a data collection program is a complex task usually involving many individuals, a time sequence of events, deployment of specialized equipment, and similar concerns. Successful implementation of an estimation technique should recognize the need for adequate distribution of samples in time and space. It should also clearly define the roles of different specialists and support staff, the timing and costs of sampling, plans for later data analysis and reporting, and custody of any samples or numerical results.

6. *Where Can I Learn More About the Ecosystem, People, and Culture of Atlantic Canada?*

The following references provide a variety of information on marine mammals, their population ecology, and their relationship with other species in the marine food web.

Nigel Bonner. 1990. *The Natural History of Seals.* Facts On File, Inc., New York.

D. H. Cushing. 1995. *Population Production and Regulation in the Sea: A Fisheries Perspective.* Cambridge University Press, New York.

J. R. Flowerdew. 1987. *Mammals: Their Reproductive Biology and Population Ecology.* E. Arnold, London, U.K.

J. A. Hutchings and R.A. Meyers. 1994. What can be learned from the collapse of a renewable resource? Atlantic cod, *Gadus marhua*, of Newfoundland and Labrador. *Canadian Journal of Fish. Aquat. Sci.* Vol. 51.

E. C. Pielou. 1974. *Population and Community Ecology: Principles and Methods.* Gordon and Breach, New York.

Randall Reeves, Brent Stewart and Stephen Leatherwood. 1992. *Seals and Sirenians.* Sierra Club Books, San Fransisco.

Marianne Reidman. 1990. *The Pinnipeds.* University of California Press, Berkeley.

David Therbune. 1973. *The Harp Seal.* Burns & MacEachern, Toronto, Canada.

Robert Ronconi compiled most of the research for this problem and contributed extensively to its writing. The International Marine Mammal Association provided a range of research materials and reviewed pre-publication drafts. The author acknowledges these contributions with gratitude.

Costa Rica

Case Study 9

Costa Rica

"How can we reduce the impacts of ecotourism on the Costa Rican environment?"

1. What Is the Background?

Ecotourism in Costa Rica

Over the past 20 years, and particularly the past decade, tourism related to natural history has been increasing steadily. This industry, often called "ecotourism", has been especially important in less developed countries, where it can have enormous economic benefits.

In Costa Rica, tourism ranks third among national sources of income, just behind banana and coffee export. Tourism has also shown a steady increase in contribution to the national economy over the period from 1979 to

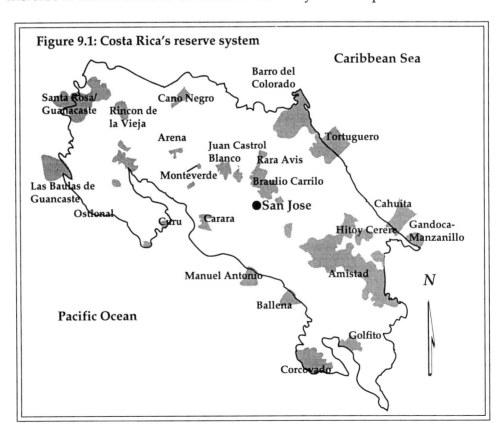

Figure 9.1: Costa Rica's reserve system

1986 (see Figure 9.2). Tourism now represents about 20% of Costa Rica's total foreign exchange.

Costa Rica's stable political condition, pleasant climate, and high level of education and health care have made it an attractive destination for ecotourists. Although the capital, San Jose, is a popular attraction, an increasing number of tourists are choosing to visit the

country's many protected areas. While most visitors to Costa Rica once came from other Central American countries, European and North American visitors now dominate the tourism market. In 1988 alone, U.S. visitors outnumbered Costa Rican visitors to a single cloud forest reserve by a factor of 4 to 1. More than a third of visitors to Costa Rica list ecotourism as a major reason for their visit.

Although ecotourism is intended to be respectful of the natural environment, and its customers more educated and responsible than the norm, unplanned and uncontrolled ecotourism can be devastating to an ecosystem. There are many examples of this, from Hawaii's intensive shoreline use, which has impacted heavily on breeding whale populations, to the mountain trails of Nepal and Peru, which are now littered with refuse from careless hikers. In Kenya, a popular safari destination, more than 1 million visitors arrived in 1990, causing observers to warn that tourists had better visit the country soon, or risk finding wild areas destroyed and overpopulated with tourists.

The Costa Rican government has taken special care in protecting valued areas from the impacts of ecotourism, including restricting camping activities and even issuing quotas on the number of visitors allowed in a park at any one time. Yet ecotourism has still taken a toll on the country's natural areas. Because restrictions on visitor activities can have important economic impacts on the tourism industry and therefore on the national economy, there is a need to balance the needs of ecotourists against conservation priorities. Revenues from ecotourism (as with other forms of tourism) can depend on external factors such as weather, political forces, and exchange rates.

2. *What Problem Are We Trying to Solve?*

Ecotourism has been identified as an important and sustainable development initiative for Costa Rica. It seems on the surface to be a good way to protect valued ecosystems while building economic capacity. Yet ecotourism will develop only if it is profitable—if the benefits of development exceed its costs.

Many issues arise in the evaluation of ecotourism. For the present, we will limit our consideration of those issues to the problem of placing a value on the Costa Rican tourism industry. With this information in hand, we can begin to weigh this industry against others and to compare the value of the industry against its environmental, political, and capital costs.

Placing an accurate value on the ecotourism industry is not a simple problem. Is the value of the industry only the fees that people pay to enter ecological reserves? Is it also the value of the hotel, restaurant, and souvenir businesses that support tourism? Is there some "value" inherent in the reserves themselves? What if the ecological reserves also support extractive uses such as rubber, fruit, or nut production? Should we add the value of those industries to the total? And so on.

In fact, the way we define the "value" of an industry, an activity, or a facility depends on the perspectives of the society doing the valuation. The point has been made in earlier chapters that these decisions are often best made by the stakeholders in the issue—the people who have a direct interest in the outcome of the decision. This is because those individuals may have different value systems, and different societal priorities, than people outside the problem. While decision-makers can certainly learn from the experience of other geographic areas, the final decision about what is most important, most highly valued, and most worthy of protection, is often a local one.

In the following discussion, we will examine the current status of the ecotourism industry in Costa Rica, including the reserve system, its supporting infrastructure such as hotels and restaurants, and current tourism demand. We will then review possible approaches to valuing this suite of activities.

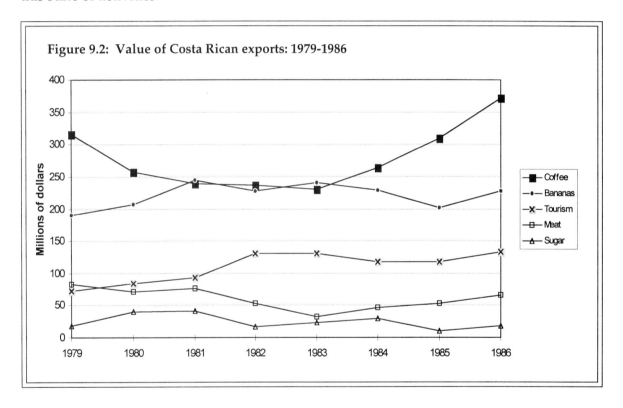

Figure 9.2: Value of Costa Rican exports: 1979-1986

3. What Components of the Environment Are Affected, and How?

Topography

Costa Rica is a small country, approximately 52,000 km² in area (about the size of Scotland), which nevertheless supports a wide variety of ecosystems. In large part, this variety is attributable to the presence of four mountain ranges and the resulting topological, altitude-related climate, and habitat variability. Mountainous regions occupy most of the central, west, and southern part of the country. To the northeast are flat, humid lowlands.

The country contains large tracts of tropical rain forest, tropical dry forests, and cloud forests, as well as a variety of other mountain habitats, mangrove swamps, black and white sand beaches and coral reef ecosystems, and volcanoes. It is these dramatic natural features that have attracted large numbers of visitors to the area.

Climate

Costa Rica is located wholly within the tropics. As much as 8 m of rain falls each year on the mountain slopes. Close to the seashore, this rainfall supports dense wet tropical forest, with trees 50 m or more in height. Rain forest continues up the mountain slopes to stunted cloud forest on the mountain peaks.

Although much of the southern and western regions of Costa Rica receive rainfall year-round, northern areas, largely dry deciduous forest, experience a dry season with leaf-fall from December to the end of March.

Costa Rica's ecological reserves

Fifty years ago, about 75% of Costa Rica was forested; today, only about 20% of the country retains that forest cover. From the mid-1960s, the country's government has tried to slow deforestation rates and protect remaining forest through a system of national parks.

At present, about 15% of the country's land area is under protection. There are 20 national parks and reserves, encompassing most of the country's habitat types and nearly all indigenous species. It is estimated that Costa Rica has about 1,500 species of trees, 850 species of birds, and more than 8,000 species of flowering plants, including 1,500 species of orchids. In fact, the forest flora is very rich in epiphytes (plants that use others as a substrate), with most surfaces colonized by some species of moss, liverwort, orchid, bromeliad, or cactus.

Although Costa Rica is known worldwide for these impressive natural history resources, many of its 19 national parks lack the infrastructure necessary to support even a basic level of tourism. Necessary infrastructure includes not only hotels and restaurants but also trained guides and visitor centers with interpretative materials.

The Costa Rican tourist industry supports four main categories of tourism: nature and adventure tourism, sun and beach tourism, cruise ship tourism, and convention or business tourism. The country's national tourism policy is currently to encourage nature and adventure tourism, especially through day trips to reserves (camping activities now being limited or prohibited in many areas). There is also an initiative to improve domestic air service and to improve roads linking the capital with the Caribbean and Pacific coasts.

> **Box 9.2: A contrast: ecotourism in the Galapagos Islands (Ecuador)**
>
> *Tourist tax: $40 U.S.*
>
> *Maximum number of visitors admitted (1974): 12,000*
>
> *Maximum number of visitors admitted (1990): 60,000*
>
> *Actual number of visitors (1994): approximately 80,000*
>
> *Emerging problems: habitat destruction, excessive water use, generation of tourism-related pollution*

The economics of tourism

There is enormous competition for land in Costa Rica, with agriculture a primary use. Where land has been taken out of the agricultural system (for example, for protection as a park), it is important to demonstrate that alternative land use is equally beneficial to the country's economy.

It has long been argued that the national park system pays for itself through tourist revenues, but in fact, although tourists brought $506 million U.S. into the country in 1993 and despite the fact that most tourists visit at least one park during their stay, the income from park visits is only about half a million U.S. dollars a year—not nearly enough to maintain the park network. This is partly because entrance fees are very low—about $2 U.S.—and because of the large number of private tour operators whose profits are not available to the park system but remain in the pockets of rich urbanites.

The Monteverde Cloud Forest Reserve is a privately run conservation reserve, established by American scientists in 1972 and now run by the Costa Rican Tropical Science Center. Although the Reserve charges $8 as an entrance fee, well above the fee charged in the national parks, and receives more than 15,000 visitors a year (making it one of the country's most popular attractions), it cannot sustain itself on its revenues. The Reserve relies heavily on donations from individuals and environmental non-government organizations. Indeed, it has been these donations that have allowed the park to expand from its original 300-ha area to its current size of more than 10,000 ha.

Another, newer park, Rara Avis, is part of the national park system. Rara Avis is much smaller than Monteverde, at only 1,300 ha, and contains both original forest and secondary growth. The park admits only 60 visitors a day. It has established two objectives for park use: tourism (deluxe accommodation in the park can cost over $90 U.S. per night) and research into sustainable extraction of forest resources. The latter includes selective logging, plant harvesting for pharmaceutical or horticultural use, raising certain butterflies for export to collectors, and farming pacas (a large rodent) for human consumption.

In a third example, Tortuguero National Park, on the country's northeast coast, experienced a 24-fold increase in visitors between 1982 and 1992. Visitors come to see sea turtle nesting and diverse inland habitats, as well as to engage in sport fishing off the coast. More than half the residents of Tortuguero, a town of about 300 people, work at least part time at the park or in neighboring research facilities. In recent years, the park has become very important in the local economy, especially since the demise of the turtling industry of 100 years ago and the more recent lumber/sawmill industry, which failed in the early 1970s.

The importance of ecotourism to the local economy, its rapid growth, and its potential for ex-

---

Box 9.3: Tour guide training: an essential element in sustainable ecotourism

As ecotourism continues to increase, it becomes especially important to ensure that ecological reserves can support the educational needs of visitors, while protecting the reserve's natural resources. This goal can be fulfilled by careful tour guide training, taking into account resource management requirements, visitor needs, and local economic conditions.

Recent studies have shown that a good tour guide program, developed in collaboration with local communities, scientists, park managers, and the hotel industry, can:

1. Help mitigate negative tourism impacts

2. Provide environmental education to adults and children in the local community

3. Enhance visitor experience by providing environmental interpretive information

4. Provide local economic benefits through lucrative part-time employment

pansion underline the importance of attaching accurate values to the industry and its adjuncts. In the following section, we'll look at one way to do this valuation.

4. How Can I Analyze This Information?

Quantifying the "value" of an activity

What is the "value" of an activity, an industry, or a facility? How can we attach dollar values to intangible factors such as aesthetics or rarity? *Should* we attach such values—or is it somehow unethical to do so?

The reality of public decision-making is that ultimately much of it is based on costs and benefits. Some legislation even requires that a formal cost-benefit analysis be conducted, to ensure that the benefits of an action outweigh its costs.

Unfortunately, it is not always easy to quantify the costs and benefits of environmental decisions. Even those who favor the use of cost-benefit analysis hotly debate how that analysis should be conducted.

It is not the purpose of this chapter to decide whether cost-benefit analysis is appropriate in evaluating ecotourism in Costa Rica. However, it may be useful for us to examine one way that analysts have used to assign dollar values to the mix of activities we call "ecotourism".

The consumer surplus

It's not difficult to think of examples of situations where we consider something to be worth more than the fee we pay to use it. For example, we may place a higher value on a skating rink located close to our home than we would on a rink located several kilometers away, even if the admission price were the same for both facilities. Similarly, people often place a high value on things like the presence of a diverse bird population, or the quietness of a rural location, relative to similar locations without those features. Environmental economists call this extra value the "consumer surplus". The problem in an analysis of environmental economics is to assign a dollar amount to this consumer surplus.

There are various ways to estimate this value. One way is to ask key informants knowledgeable about local conditions to estimate the value. Another is to conduct a survey of local residents, for example asking what they would be willing to pay to have that attribute continue; this approach is called "contingent value" analysis. Various qualitative methods also exist, including ranking or weighting of various attributes to gain insight into their relative values.

An example: the travel-cost method

One method that has been used to value recreational opportunities to which people must travel is called the travel-cost method. This method is based on the principle that as prices rise, people purchase less of a consumer good. We can extend this idea to the purchase of ecotourism trips.

Cost-benefit analysis was developed by the U.S. Corps of Engineers in the mid-1930s as a way of evaluating the impacts of dam projects. The technique has received considerable criticism over the years, primarily because the choice of a discount rate (the rate at which the value of a "dollar" changes over time, including interest rate changes, inflation, and similar factors) has such a strong influence on the outcome of the analysis, and because of the difficulty of valuing "intangibles" such as human health and wildlife habitat. Although techniques are available for estimating the "value" of these elements, people apply those techniques in widely varying ways and thus come to widely varying conclusions.

There is also considerable controversy about what boundaries should be placed on the analysis. For example, in estimating "benefits" should we include not only the jobs created in environmental clean-up but also the spin-off benefits of those jobs, such as increased purchasing of homes, cars, and consumer goods by those employed in the clean-up. It's often very difficult to decide which items should be included and which excluded from the analysis.

Simply stated, the travel-cost method looks at the distance traveled by the consumer, including not only actual travel costs (gasoline, wear and tear, etc.) but also any out-of-pocket expenses (tolls, snacks, etc.) and the value of the individual's time in undertaking the travel, and use these amounts to estimate the value of the travel experience to the consumer.

In a recent examination of travel costs at Monteverde Cloud Forest Biological Reserve, Dave Tobias and Robert Mendelsohn of the Yale School of Forestry based their estimate on three key measures: distance traveled (estimated at a cost within Costa Rica of about $0.15 U.S. per km), local population density in the canton (province) of the tourist's origin, and illiteracy rate. Other combinations of variables would be possible, but Tobias and Mendelsohn believed these to be the most important factors. Their reasoning was based on the observation that where population density is higher, people will tend to travel out of the area more frequently, while people who live in relatively unpopulated areas are likelier to have unspoiled areas nearby. Similarly, the higher the illiteracy level, the lower the likelihood that people will be interested in visiting a remote ecological reserve, while high visitation rates are correlated with higher levels of education (and probably permanent income).

How should we interpret these results?

The travel-cost method gives us one way of estimating the true "value" of a complex resource, in this case the Monteverde Cloud Forest Biological Reserve. It should be emphasized that the value we arrive at in an analysis such as this is simply an estimate, not a fact. Obviously, Tobias and Mendelsohn's analysis was based on a number of assumptions regarding the cost of travel within Costa Rica. They intentionally ignore other potential "values" of the cloud forest, such as extractive uses and harvests. And they have chosen a 4% real interest rate, which may or may not appear realistic now or in the future.

Do these considerations make these results unusable? Not at all.

Tobias and Mendelsohn use their approach to draw broad conclusions about the advisability of expanding the reserve into neighboring lands. These sorts of conclusions may not be significantly affected by changes in travel cost or even in interest rate. Secondly, *any* method we use to estimate the "value" of this complex resource will have weaknesses. Our task, as environmental problem-solvers, is to understand those weaknesses and their potential impact on the decision we are trying to make. This means thinking critically about the components of the analysis, the assumptions that underlie it, and the conclusions that have been or could be drawn from it.

The sorts of questions that arise in reviewing a travel-cost method analysis are part of the reason that decision-makers often use several methods to analyze a single situation. If a major decision is to be made (often, if a lot of money is to be spent), the careful analyst will examine the problem from several perspectives, of which travel-cost analysis might be only one. Several analyses taken together, particularly if they are supported with information about potential error, can provide a powerful suite of tools for even the toughest environmental decision.

The problem of contradictory and incomplete results

Performing your first analysis of an environmental problem can be quite an eye-opener. It comes as a surprise to many students that the data you need to answer an environmental question are often incomplete or totally unavailable. Worse, it is not at all uncommon to have different experiments, surveys, or other data-collection exercises result in information that contradicts the findings of other studies.

Learning to manage this problem is an important skill. There are several challenges in this.

Filling data gaps

One approach to filling in missing information is to make reasoned guesses and document them in your analysis. If you adopt this approach, it is important that you be able to describe and quantify the error implicit in your guesses.

If Tobias and Mendelsohn had complete and accurate data for Monteverde, but not for neighboring parks, they could still draw conclusions about those parks from the Monteverde results. But they would not assume the results for those other areas to be the *same* as for Monteverde. Rather, they would try to understand how those areas differ from the area where they gathered their original data and then try to estimate how much the differences would affect the outcome of their analysis. In a good analysis, you will always see an estimate of the "error," or uncertainty, of the conclusions.

Collecting new data

Sometimes, there are simply not enough data available even to begin an analysis. In this case, it may be necessary to collect new information through field surveys or other techniques.

In some cases, it is possible to use historic data from a site as the basis for a new data-collection program—in other words, to extend the available data base with new information. Case Study 8 discusses some of the issues that arise in the collection of a good data set.

Resolving conflicting results

Probably the toughest task for the data analyst is to decide which of two or more contradictory data sets is "correct." There is no easy solution to this problem—in fact, debates about data quality rage throughout the scientific literature. In her book *Science Under Siege*, Beth Savan discusses a number of such debates. Box 9.6 summarizes her analysis of one of these.

Box 9.6: The great lead debate: an example of contradictory conclusions from a single data set

One of the most prominent recent debates about data quality revolved around historical levels of lead in the atmosphere. If it can be proved that preindustrial levels of lead in air were much lower than those we observe today, we can begin to build a case that current lead levels have arisen from human activities such as vehicular emissions, incineration, and smelting. If, on the other hand, lead levels in the past were similar to current levels, then we need not be so concerned about current anthropogenic emissions. The economic implications of either conclusion are enormous, because they affect a wide range of human activities and industries.

One of the key data bases in this debate comes from cores of glacial ice. The ice, which has not melted in many hundreds or thousands of years, contains trace amounts of materials present in the atmosphere at the time the ice was formed. Among these materials is lead. The age of a core sample can be deduced from the depth of the ice where the sample was taken—the deeper the sample, the older it will be.

In a highly emotional debate, two groups of scientists arrived at radically different conclusions about lead concentrations in these samples. Scientists from the Polish Central Laboratory for Radiological Protection published a report in the highly credible scientific journal **Geochimica et Cosmochimica Acta** *observing that levels in the samples they collected were similar to those observed in the modern world. They therefore concluded that the rate of metal deposition had not changed with the advent of industrial society.*

By contrast, Clair C. Patterson, a respected analytical geochemist, has argued fiercely that the samples were contaminated with modern-day air when they were collected, thus artificially raising the concentrations above what would actually have existed in the past. He believed that the sample-collection methods and the laboratory practices used to analyze the cores allowed contamination to creep in at several stages. He concluded that the data set is "not trustworthy" and the findings of the Polish scientists are "erroneous".

Who is correct? Ultimately, the analyst must decide for himself or herself which conclusion is most persuasive. In view of the social, economic and political implications of many environmental decisions, it is especially important that the analyst be able to justify his or her conclusions fully and persuasively.

5. How Can I Use My Findings to Reach a Solution?

Use the decision-making framework described in the Introduction to organize your thinking on this problem, as follows:

1. What is the problem?

In Section 2, we identified the problem as placing a value on the ecotourism industry in Costa Rica, including the reserve system, its supporting infrastructure such as hotels and restaurants, and current tourism demand.

2. In what ways do human activities have impact on the natural environment to cause "a problem"? How do these mechanisms give you clues to possible solutions?

This case study describes the direct impacts that humans have on Costa Rica's rain forests, for example through trampling and litter, and also the indirect impacts, for instance on local businesses and attitudes toward the environment. This suggests to us that an ideal approach to valuing ecotourism should encompass not only obvious, direct effects, but also those that are indirect or secondary. In other words, it's not enough to examine only the dollars flowing into the country from tourism activities, even if we include the "costs" of environmental impact. We must also consider the positive (revenues) and negative (costs) effects of a wide range of activities such as training programs and increased local awareness. The method described in this case does not achieve all of this, but does illustrate an application that goes beyond the obvious "costs" of ecotourism.

3. What governments are responsible for the issue? Whose laws may apply?

The Costa Rican federal government is central in this case, although municipal governments may become involved in the development and control of specific projects.

4. Who has a stake in the problem? Who should be involved in making decisions?

Other than the governments listed in step 3, possible stakeholders/decision makers would be local residents, who are also potential park and hospitality workers. Hotels and restaurant owners and the owners of souvenir shops would have a strong interest in the issue. Environmental non-government organizations such as the World Wildlife Fund are keenly interested in preserving Costa Rica's rain forests. Finally, domestic and international tourists intending to visit the reserves would be important stakeholders.

5. In the view of your decision-making group, what are the attributes of a satisfactory solution? In other words, when will you be satisfied that the problem is "solved"?

A satisfactory solution is one that yields a comprehensive and accurate value for Costa Rican ecotourism. The extent to which any value reflects a comprehensive and/or accurate analysis can only be judged by the decision-making group. A solution that one group believes to be comprehensive could be rejected by another group as incomplete.

6.   How will you evaluate (test, compare) potential solutions?

Environmental economists continually debate what should, and should not, form part of an analysis. To a large extent, the choice of a "best" method is a subjective one based on the needs and priorities of the decision-making group (and see step 5, above). The important thing is that the analyst be able to justify the choice of method and defend the components analyzed. There may well continue to be disagreement about whether the chosen method is adequate for the task at hand. It is therefore essential that the selection of an analytical method reflect the consensus of the decision-making group and, ideally, the full range of stakeholder opinion. Sometimes, this means using more than one method to reach a decision about "value".

7.   What are all the feasible solutions to the problem?

There is a large literature on valuing intangibles, assessing the consumer surplus, and so on. Many methods are available, some of which are discussed in this case. Others can be identified through a more intensive literature search, or the analyst is free to develop—and justify!—new analytical approaches.

8.   Which solutions work "best" in terms of the attributes you identified in (5)?

As discussed in steps 5 and 6 above, the "best" solution will depend on the needs and priorities of the decision-making group. As a very simple example, if only park managers were involved in the decision, they might be satisfied with a method that assessed the capital costs of building visitor centers and restrooms, the costs of hiring staff, and the revenues to be gained from visitors. They would likely be less concerned about travel cost, for example, or the overall "value" of the ecotourism industry to the Costa Rican economy.

9.   Which solution will be easiest to implement?

As in Case Study 8, the practical implications of any decision about method will include whether necessary resources (for instance computers) are available, whether staff have adequate training to conduct the analysis, whether adequate documentation of the method is available, and similar concerns. Methods such as that described in Section 4 are typically applied by university and government researchers or by consultants. They are seldom used by tourism operators or visitors.

10.   What steps are needed for successful implementation? Who will pay? Who will monitor progress?

This answer to this question will depend on the chosen method and the group responsible for conducting it. Consideration should be given to the costs of analysis, and who will pay those costs. Timing and duration of the analysis could also be defined, including (if appropriate) a schedule for periodic reanalysis of the data. Custody of data, verification of analytical results, and reporting responsibilities should also be defined as part of an implementation plan.

6. Where Can I Learn More About the Ecosystem, People, and Culture of Costa Rica?

The following sources provide a variety of background information on Costa Rica's people and environment, and on ecotourism in Costa Rica and elsewhere.

E. Boo. (undated). Ecotourism: the potentials and pitfalls. World Wildlife Fund, Washington, D.C.

D. Burnie. 1994. Ecotourists to paradise. *New Scientist*, 16 April 1994.

P. Dabrowski. 1994. Tourism for conservation, conservation for tourism. *Unasylva* 176(45): 42-44.

S. Jacobson and R. Robles. 1992. Profile: Ecotourism, sustainable development, and conservation education: development of a tour guide training program in Tortuguero, Costa Rica. *Environmental Management* 16(6): 701-713.

J. Laird. 1993. Laos pins tourism hopes on unspoiled nature and culture. *Our Planet* 5(4): 8-11.

Organisation for Economic Co-operation and Development. 1993?. Tourism policy and international Tourism in OECD Countries 1991-1992. OECD, Paris, France.

B. Savan. 1988. *Science Under Siege: The Myth of Objectivity in Scientific Research.* CBC Enterprises, Toronto, Canada.

D. Tobias and R. Mendelsohn. 1991. Valuing ecotourism in a tropical rain-forest reserve. *Ambio* 20(2): 91-93.

The following are the key references in the "great lead debate".

Z. Jaworowsky, M. Bysiek, and L. Kownacka. 1981. Flow of metals into the global atmosphere. *Geochimica et Cosmochimica Acta* 45(1981), pp. 2185-86.

C. C. Patterson. 1983. Criticism of "Flow of metals into the global atmosphere." *Geochimica et Cosmochimica Acta* 47(1983), p. 1163.

Case Study 10

Sri Lanka

"How can we increase production while decreasing soil erosion in an agricultural watershed?"

1. What Is the Background?

A rich agricultural tradition

The Republic of Sri Lanka is located off the southeast coast of India, in the Indian Ocean. It is a country small in area—only about 66,000 km², but rich in agricultural resources. Most of the population of 18,000,000 live in rural areas, with a population density of about 225 people per km². The country supports a number of ethnic groups and religions (and has thus sometimes been the focus of religious conflict) and has a high rate of adult literacy (> 85%) and a relatively long life expectancy (>66 years) compared to many less developed countries. It now ranks high among developing and developed countries in terms of several such key human development indicators.

More than half the population is employed in agriculture, especially in the production of rubber, tea, and coconuts in the country's wet south and southeast regions. These products are in fact the country's main exports, to countries such as Japan, India, the United States, the United Kingdom, and China.

In ancient times, Sri Lanka was considered to be the center of Asian rice production. Its rice-farming practices depended on an

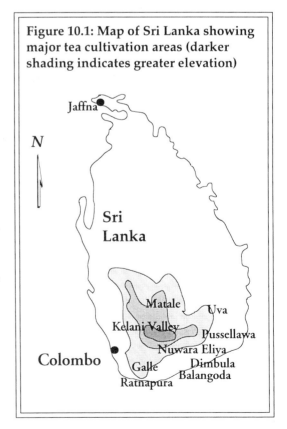

Figure 10.1: Map of Sri Lanka showing major tea cultivation areas (darker shading indicates greater elevation)

elaborate system of reservoirs, irrigation channels, and terraces. Although rice continues to be the staple food crop of the population, it is now mostly imported from China, Sri Lanka's principal trading partner, and from Thailand and Pakistan.

Today, one of Sri Lanka's most important crops is tea, which earns more than 66% of the country's foreign currency. Sri Lanka, once known as Ceylon, is renowned for its beautiful highland areas, warm climate, and abundant rainfall. These qualities also make it ideal for raising the tea plant. Most tea is produced in the country's southern highland areas, at 650 m or more above sea level, on level terrain with a hot, moist climate. Teas are classified by their growth altitude: those grown over 1,000 m are categorized as Highgrown; those grown between 600 and 1,000 m as Medium-Grown; and those grown below 600 m as Low-Grown. Altitude is a more reliable indicator of quality than is district, since a single mountainous district may produce tea at altitudes differing by more than 1,000 m.

Tea production in Sri Lanka dates mostly from the late 19th century, when tea plantations were developed by British settlers to serve the thirsty British market for tea. Other former British colonies such as Kenya and India also support major tea planta-

> **Box 10.1: A brief introduction to Sri Lankan tea**
>
> *Most tea consumed in North America, Europe, and Australia is blended. Even "Ceylon tea" is often a blend of different Sri Lankan teas. Here is a brief summary of some important tea varieties:*
>
> ***"Black tea"*** *is produced in the high altitudes of southeastern Sri Lanka. This type of tea is made from dried, fermented tea leaves. Its infusion is rich in flavor and aroma and brews to a light golden color. Four main varieties, named for their plantation areas, are recognized in Sri Lanka: Nuwara Eliya (which some consider the "champagne" of Sri Lankan tea), Uva (more strongly flavored), Dimbula, and Kandy.*
>
> ***"Green tea"*** *comprises about 20% of world tea production and is mostly produced in China, Japan, and Taiwan. It is made from steamed tea leaves, a process that allows the leaf to retain its green color. This variety is more subtle in flavor and aroma and brews to a pale yellow-green color. It is sometimes flavored with jasmine and sold as jasmine tea.*
>
> ***"Oolong tea"*** *is made from partially fermented leaves. Its flavor is intermediate between black and green tea.*

tions, and indeed mixtures of Sri Lankan ("Ceylon"), Indian, and Kenyan teas now comprise the majority of tea sold commercially in North America.

Today, textiles are beginning to overtake tea as the island's principal export. Twenty years of import-substitution policies have created an economic climate that is biased against export-oriented agriculture. In addition, declining international prices for tea have discouraged plantation owners from replanting their aging plants. As a result, tea planters now have falling productivity and cannot as easily accommodate the economic impacts of international price cycles.

Until 1992, tea plantations in Sri Lanka were owned by two state corporations employing more than 425,000 workers. With declining yields and limited investment in replanting and maintenance, however, the government has contracted with private sector businesses for the management of its tea operations. Twenty-two private sector companies now manage 449 tea estates on a profit-share basis. These and other reforms should increase the profitability of the tea industry in Sri Lanka, although heavy national emphasis on industrial development now places enormous pressure on the agricultural industry, especially for land. All good tea lands have already been developed; the search for new lands may take planters into marginal and environmentally fragile lands.

2. *What Problem Are We Trying to Solve?*

The challenge of sustainable agriculture

Although Sri Lanka is beginning to diversify its agricultural base and to extend its industrial base, tea remains a crop of great economic importance to the country. The production of tea has been declining steadily since about 1965, partly because of reduced land area under cultivation (some land having been diverted to urban development) and partly because of reduced productivity in the lands under cultivation.

Several factors have contributed to reduced productivity. In recent decades, tea production has moved into steeply sloping territory, away from the flatter "plains" areas that were planted many decades ago. On these steeper slopes, soil management practices have often been careless and this, along with assiduous weeding, has created serious soil erosion problems that in some cases have permanently affected soil quality and thus tea yield.

In an effort to extend planting areas, some farmers have cleared away the large trees that formerly shaded the tree plants. In doing so, they not only promote soil erosion but also adversely affect the quality of tea grown: leaves are larger and there are more of them in sunny areas, but the flavor of these leaves is not as good and they are less desired in the marketplace than those grown more slowly, in more shaded areas.

Finally, although new high-yielding vegetatively propagated (HYV) tea varieties have been available since the 1960s, most tea under cultivation is still "seedling tea"—older varieties yielding only about one-third of the yield possible with HYV tea (850 to 900 kg/ha vs. HYV's 2,500 kg/ha).

The problem we face in this case is therefore essentially one of economically and enviromentally sustainable agriculture. How can Sri Lanka grow good tea, slowly, while preserving the land base for future tea crops?

Box 10.2: Sri Lanka's famous "seasonal" teas

Part of Sri Lanka's fame as a tea-producing nation comes from its so-called "seasonal teas". These teas, which are produced in relatively small amounts but which are highly prized by buyers, are grown only in certain areas of the country, particularly Dimbulla/Dickoya and Uva, and only over a few months of each year when climatic conditions are suitable.

These "seasonal" teas are used in small quantities to impart an unmistakable character— "quality"—to tea blends. To the discerning consumer, they are almost irreplaceable. If the quantity or quality of these teas declines, or becomes erratic, or if they become too rare and therefore too costly, tea packers will be forced to replace them, thus changing the taste of the blend and potentially reducing its appeal for consumers. In 1980, the Dimbula/Dickoya season failed, and in 1981 the Uva season was equally unsuccessful. Since that time, Sri Lanka has had to work hard to regain the confidence of packers so as to avoid decreased popularity and demand for these special teas. Tea crop failures have been partly to blame for the troubled condition of the tea industry and the push to increase productivity to guarantee a good crop.

---

3. What Components of the Environment Are Affected, and How?

Climate and land use

Although Sri Lanka has a relatively small land base, it supports a wide range of agro-environmental conditions. In the mid 1980s, about 2.2 million ha of land was used for permanent cropping; of this, about 0.24 million ha was planted in tea. (Almost half the country's total cultivated area is in shifting cultivation, planted to a variety of crops. The remainder is in rubber, coconut, rice, mixed crops, and minor export crops such as spices.)

The country has three major climatic zones—wet, intermediate, and dry—based on rainfall, vegetation, soils, and land use. Within the wet and intermediate zones, there are subdivisions of climate based on elevation.

A very important climatic force in Sri Lanka, as in much of Asia, is the monsoon. The heavy monsoon rains hit the western part of the island from July to September, but Uva and other eastern parts of the country are protected from the rains in this period and experience dry winds suitable for the best tea production. From December to March, the monsoon moves from the northwest to hit the eastern part of the island. The high moisture levels promote good tea growth but poorer flavor; during this period, southern and western regions including Dimula and Dickoya experience the dry winds necessary for their "season" of tea growth. Nuwara Eliya, located on a mountain ridge running north-south through the island, produces an excellent tea year-round.

In the wettest parts of the country, mean annual rainfall can exceed 5,000 mm. By contrast, semihumid regions of the country receive only 1,250 mm of rain a year. The heaviest rains occur during the monsoons. Between the monsoons, most regions experience occasional short, intense convection thunderstorms, in which as much as 100 mm of rain can fall in a single hour.

Cultivation practices

Tea plantations are created out of jungle. First, the dense undergrowth is cut and burned, then large trees are felled and burned or removed. Grassland is burned off. When the land is clear, the tea plants are planted, usually in squares or equilateral triangles, with an average planting density of about 1,200 shrubs per ha. Young tea plants are not picked for 3 years, at which time the "flush," or first growth, is harvested. Well-managed tea plants can live more than 50 years.

Where tea is planted on hot, moist plains, with suitable climatic conditions, leaf yield is high. At higher altitudes, steeper slopes must be terraced for adequate cultivation, the growing season is usually shorter, and the plants are more difficult to pick. The teas from these areas tend to have lower yields but better flavors than those grown on the plains. Nevertheless, it is these upland areas that are most vulnerable to soil erosion and surface runoff, especially since these valued plantation areas are located near the headwaters of major streams.

Older tea-planting practices usually involved clearing long, sometimes steep slopes and planting up and down the slope, parallel to the direction of flow. The spaces between the rows of tea are weeded thoroughly and regularly, exposing the soil to the impact of rain and also loosening a layer of soil that can be easily transported by runoff water. Although drains are sometimes built to divert rainwater, they are seldom planted or paved, and are thus themselves subject to erosion.

In the wild, a tea plant is really shrub growing to 9 m or more in height. In cultivation, tea plants are pruned to about 1 m high. Plants take 3 to 5 years to mature. Mature plants are picked ("plucked") by removing the new growth, or "flush"—several leaves and a bud. Fast-growing plants, for instance at lower altitudes with abundant moisture, produce a flush every week or so through the growing season. At high altitudes, it takes 2 to 3 weeks to produce a flush.

As a tea plantation ages, individual plants die and a gap opens in the canopy. These gaps are rarely replanted, either with tea or with some other plant. Older plantations may therefore be at particular risk of soil erosion, and there is evidence that as much as 43,000 ha of old tea plantation has already experienced severe soil degradation as a result of these management practices.

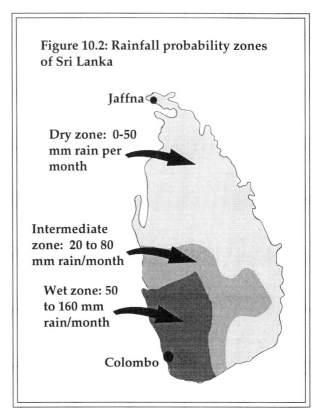

Figure 10.2: Rainfall probability zones of Sri Lanka

Jaffna

Dry zone: 0-50 mm rain per month

Intermediate zone: 20 to 80 mm rain/month

Wet zone: 50 to 160 mm rain/month

Colombo

Tea is mainly grown on Ultisol soils, a relatively stable soil type. Even though the intensity of rainfall diminishes with altitude, and Ultisols have relatively low "erodibility", severe soil erosion can occur on tea plantations because of the long, steep slopes, orientation of planting, and large exposed soil areas.

4. How Can I Analyze This Information?

Although this case describes the production of a specialty crop in a unique environmental setting, the tools we can use to resolve it, and the lessons we can draw from it, are common to many agricultural and rural nonagricultural situations.

Do we plan for tea or some other crop?

In this case, we have established that although tea can be grown in many places in Sri Lanka, excellent tea, the tea for which Sri Lanka has become justly famous, grows best on steep, high-altitude slopes.

Is it even possible to cultivate such a slope without devastating soil erosion? Certainly traditional tea-planting practices, developed for plains areas, are inappropriate for steep slopes. And certainly the nature of the tea plant, and the gaps that can occur in a plantation through plant death, do little to retard soil erosion and may often in fact encourage it.

Other crops are possible on these slopes: rubber, coconut, coffee, and spices have all been grown with some degree of success. A variety of vegetable crops are also grown, more typically in shifting cultivation. But tea is important, perhaps essential, to Sri Lanka's economy, so in the eyes of many observers the solution lies not in replacing tea with some less valuable commodity. Rather, the answer lies in implementing remedial measures to halt erosion and ongoing land management practices to prevent or minimize new erosion. What tools are available for this?

The physics of soil erosion

A single rain drop, particularly in intense rainfall events, can carry considerable energy. That energy is transferred to the soil surface on impact. Depending on the energy of the rain drop and the erodibility of the soil, a particle may become dislodged and suspended in the surface flow. The inertia of the particle is the most difficult to overcome: once suspended, it can remain in motion for some time, until the flow velocity drops to a point where the particle is able to settle out. The settled particle can once again be dislodged and transported, "leapfrogging" in this way down the slope into the nearest stream.

To reduce soil erosion on an agricultural field, we can manipulate one or more of several variables.

Common erosion control techniques

Plant a full canopy

First, we can encourage the dissipation of raindrop energy by planting a full "canopy" of plants on the field. When the raindrops fall on the field, they will fall first on the leaves of the plants, transferring much of their energy to harmless vibration of the plant stem and leaves. The rain that reaches the soil will have lost much of its ability to erode and will therefore have a greater chance of infiltrating rather than running off the field.

Increase soil-surface roughness

Another way to reduce erosion is to reduce the velocity of runoff water flowing over the field. This can be achieved by increasing the roughness of the soil surface, for instance by interplanting with a cover crop or grass, by mulching (grass cuttings have proven very effective on tea plantations), or by similar measures.

Encourage slower drainage

Field experimentation reveals that erosion will occur even on relatively stable soils if there is a long slope length and rapid water movement. Traditional planting methods encourage this long slope length by arranging rows parallel to the direction of flow. A measure that can be implemented when replanting, or planting new fields, is simply to orient the rows perpendicular to the direction of flow. Mulches and cover crops also assist in reducing flow velocity. Vegetated or gravelled drainage ditches also increase surface roughness and thus reduce runoff velocity. Stone terraces collect and slow flowing water and allow more gradual infiltration.

Trap eroded soil on the field

Even the best-managed fields will be subject to erosion under some conditions. Under these "worst case" conditions, eroded sediment can be trapped on the field by the use of buffer strips—bands of shrubs and herbaceous vegetation bordering the fields, and especially along the banks of receiving waters. Buffer strips can also help to reduce wind erosion in areas where winds have sufficient energy to lift soil from the field surface. Nitrogen-fixing shrubs such as *Gliricidia* can be used in buffer strips, providing green manure, mulch, fuelwood, and shade.

Fill vacant areas

Vacant areas on the field may be particularly at risk for soil erosion. These vacancies can be planted with grasses or other cover crops with the potential to increase the nutrient content of the soil. Large vacancies, such as those that occur in large-scale replanting, can be mulched with cut grasses.

Weed selectively

Aggressive weeding has been shown to loosen soil and encourage erosion. Newer weeding techniques focus on removing plants that are likely to compete with the tea crop while leaving more innocuous species.

Table 10.1: Mean annual rainfall, erosivity and elevation for selected tea-growing areas in Sri Lanka (1985 data)

Station	Mean annual rainfall (mm)	Erosive rain (% >25 mm/hr)	Elevation (m)	Mean annual erosivity (ft-T/A)
Galle	2,275	62	21	56,100
Ratnapura	3,200	56	40	70,600
Katugastota	1,975	47	457	36,100
Badulla	1,825	27	677	19,500
Watawala	4,000	33	994	51,300
Nuwara Eliya	1,725	4	1,881	2,700

Source: Krishnarajah 1985

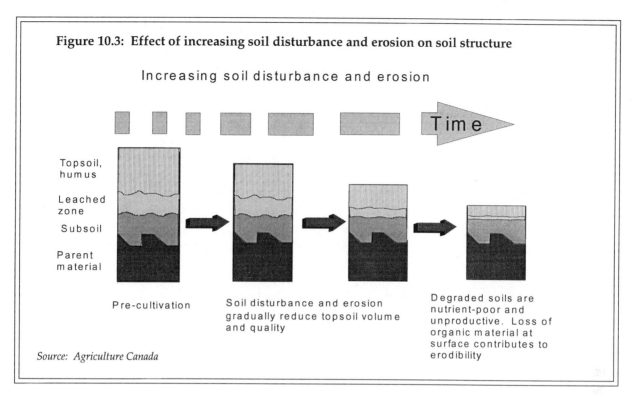

Figure 10.3: Effect of increasing soil disturbance and erosion on soil structure

Increasing soil disturbance and erosion

Time

Topsoil, humus

Leached zone

Subsoil

Parent material

Pre-cultivation

Soil disturbance and erosion gradually reduce topsoil volume and quality

Degraded soils are nutrient-poor and unproductive. Loss of organic material at surface contributes to erodibility

Source: Agriculture Canada

Diversify crops

High-value tree crops such as cloves, nutmeg, coffee, and pepper can be grown with tea or can replace tea in marginal or unprofitable plantations.

Use traditional conservation methods

Mechanical soil conservation technologies such as lock-and-spill drains, gully rehabilitation, and stabilization of streambanks and roads cannot alone resolve the problems of soil erosion on steep slopes. Traditional physical methods such as terracing, and traditional cropping and weeding approaches, may have considerable value and higher implementability than "high-tech," high-maintenance technologies.

Use an integrated approach for planning land-use and conservation strategies

Each plantation area has its own unique combination of slope, soil type, climate, and plant age, health, and orientation. Each plantation must therefore have its own strategy for land use and soil conservation. In general, the plan should emphasize maintenance of a good vegetative cover, some diversification of the cropping base (both for economic reasons and for physical diversity), a combination of traditional and modern conservation and land-use methods, and the safe disposal of runoff to stable channels and waterways.

Monitor the results

It is not enough to complete a plan and implement some conservation measures. With highly vulnerable soils, high-intensity rainfall, and a highly valuable crop such as seasonal tea, it is important to monitor the effectiveness of conservation measures by measuring streamflow and soil loss on a regular basis. Such a program can give an early warning of incipient problems and allow corrective measures to be taken promptly.

5. How Can I Use My Findings to Reach a Solution?

The development of a management plan for tea production incorporates both policy decisions—grow tea or mixed crops, for example— and technical decisions, such as which tillage practices or drainage structures are required.

1.   *What is the problem?*

In Section 2, we identified the problem as "how can Sri Lanka grow good tea, slowly, while preserving the land base for future tea crops?"

2.   *In what ways do human activities have impact on the natural environment to cause "a problem"? How do these mechanisms give you clues to possible solutions?*

In this case, we observe the combination of a fragile ecosystem—steep slopes with erodible soils exposed to heavy rainfall—and intense agricultural activity. Humans impact this environment in several ways: by cutting and clearing the jungle, by cultivating the soil, by introducing a single species of plant and constraining the growth of individual plants, and by leaving open gaps in the planted area. In the natural jungle, the physical diversity of plants and terrain reduce the energy of falling raindrops and thus their erosive potential. In a cultivated field, much more of the soil is exposed to the elements. This suggests that good remedial approaches would (a) reduce the energy of falling rain, for instance by creating a full canopy composed of close or mixed-species planting; (b) prevent the movement of soil from the field, for instance by physical barriers such as trenches and berms, or by intermixing plant species to trap eroded soil; and (c) reduce the potential for rapid runoff through the cultivated area, for instance by avoiding steep slopes, by providing a mulch, or by constructing proper drainage channels.

3.   *What governments are responsible for the issue? Whose laws may apply?*

In this case, the Sri Lankan federal government is probably the key group, because of the implications for trade. Provincial and local agencies may also have a role, however, especially in technology transfer—teaching planters better land management practices.

4.   *Who has a stake in the problem? Who should be involved in making decisions?*

Important stakeholders in this case would include the federal government, the tea planters and tea packers, and consumers of tea and alternative crops such as coffee and spices that might be interplanted with tea. Technology producers from Sri Lanka or other countries might have an interest in selling equipment such as pumps and pipes, depending on the remedial approach taken. Plantation employees would be interested, both because of possible impacts on their wages and because of changes that may be required in cultivation methods.

5.   *In the view of your decision-making group, what are the attributes of a satisfactory solution? In other words, when will you be satisfied that the problem is "solved"?*

A satisfactory solution would be one that allows planters to produce a good quantity of high-quality tea, while preserving the land base for future tea crops. It is difficult to set targets for "satisfactory" production levels or "sustainable" cultivation. (This problem also occurs in the assessment of watershed management approaches in Hubbard Brook, New Hampshire, Case Study 11.) We could set targets for reducing soil loss over a given area, for example a reduction of 100 kg/ha by 2005. Or we could measure sustainability by the amount of tea produced per ha, aiming to keep a constant level of production over the next 10 years with no change in inputs (fertilizer and pesticides, for example). It is important that targets be specific and measurable, not vague like "sustainability", or progress will be difficult to assess. The cost of management options may also be a factor, although it is probably of more interest to plantation operators than it is to government staff.

6.   How will you evaluate (test, compare) potential solutions?

Computer simulation models could be very useful here; a number of these are described in the literature, many available at no cost from public agencies. Another approach would be to use empirical evidence to estimate the effectiveness of individual remedial measures —or perhaps a range of effectiveness. Again, there is much written on the performance of measures such as those described in this case. We could then compare management options on this basis. Comparison need not be quantitative or exact as long as it allows you to separate "better" options from "worse" options. Quantitative comparison—such as that provided by simulation models—may be more important if you are trying to estimate the precise tonnage of soil lost under a given management regime.

7.   What are all the feasible solutions to the problem?

Section 4 describes a number of management options, both traditional and "high-tech", that have been used in cases like this around the world. The literature on agricultural non-point source pollution provides a rich source of information on these and other technologies. You could also contact local extension staff in your area for advice about possible management approaches. It is not usually possible to use every option in every situation, so you should be careful to find out the limitations of each approach.

8.   Which solutions work "best" in terms of the attributes you identified in (5)?

Using the evaluation method from step 6, determine the performance of each feasible management approach in terms of the targets you set out in step 5. As discussed in other cases, you may wish to assign weights to individual decision criteria such as cost or tonnage reduced. Set up your results in a decision matrix, with criteria along one axis and options along the other. Fill in the body of the matrix with scores or ranks, calculate weighted scores if appropriate, and determine the "best" option (or combination of options) from the total score (across all criteria) for each option.

9.   Which solution will be easiest to implement?

Ease of implementation will be a critical consideration in this case. There is considerable economic incentive for planters to avoid extra expenditure and simply continue with inappropriate land management practices , moving into new, if marginal, areas when production begins to fall. Convincing planters of the merits of change may require government subsidies or other incentives (for instance tax relief or preferred-supplier status). The recommended approach *must* be attractive (not just acceptable) to plantation operators or change simply will not occur. (Private operators of tea plantations work on a profit-share basis, so financial considerations may be especially important to them.)

10.   *What steps are needed for successful implementation? Who will pay? Who will monitor progress?*

There is considerable experience around the world in implementing land management changes. Much of this experience lies with agricultural extension specialists, whose job it is to provide technical assistance to farmers. There are two benefits of this experience for the environmental manager. First, most kinds of implementation problems have been encountered—and overcome—at one time or another, so there is also a good literature on implementation needs. And second, there is an enormous diversity of extension materials available for the purposes of supporting and encouraging implementation. Local extension agencies would be a good place to start in planning implementation steps, costs, and responsibilities.

5. *Where Can I Learn More About the Ecosystem, People, and Culture of Sri Lanka?*

Literature on tea cultivation is not easy to find in the average library, although it is often available on interlibrary loan from major agricultural libraries (e.g., U.S. Department of Agriculture, Agriculture and Agri-Food Canada). Literature on soil conservation is much more widely available throughout the world. The economics of the tea industry in Sri Lanka are described in a variety of publications in scholarly and popular literature.

R. L. de Silva. 1982. The production and marketing of quality Sri Lanka tea. *Tea Quarterly* 51(3): 137-139.

W. Gooneratne and D. Wesumperuma. 1984. *Plantation Agriculture in Sri Lanka: Issues in Employment and Development*. International Labour Organisation, Asian Development Programme, Bangkok, Thailand.

J. S. Gunasekera. 1991. Integration of traditional and modern conservation methods on hillslopes in Sri Lanka. In: W. C. Moldenhauer, N. W. Hudson, T. C. Sheng, and San-Wei Lee (eds.), *Development of Conservation Farming on Hillslopes*. Soil and Water Conservation Society, Ankeny, Iowa.

P. Krishnarajah. 1985. Soil erosion control measures for tea land in Sri Lanka. *Sri Lanka Journal of Tea Science* 54(2): 91-101.

C. Rodrigo. 1982. The view-point of the producer. *Tea Quarterly* 51(3): 142-144.

E. Thorbecke and J. Svejnar. 1987. Economic Policies and Agricultural Performance in Sri Lanka 1960-1984. Development Centre Studies, Organisation for Economic Co-operation and Development, OECD, Paris, France.

U.S. Agency for International Development. 1986. Formulating agricultural policy in a complex institutional environment: the case of Sri Lanka. U.S. Agency for International Development, Bureau for Science and Technology, Office of Agriculture.

Hubbard Brook

"What is the best way to log a second-growth forest?"

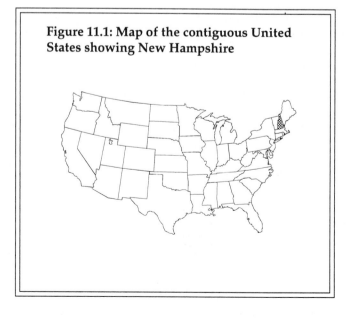

1. What Is the Background?

A complex network

A forested ecosystem is a complex network of interactions between living and nonliving components, including energy flow, materials cycling, and physical, chemical, and biological processes. We do not currently understand these processes very well, and as a result resource managers often have difficulty assessing the impacts of human activities on the ecosystem they are "managing". Secondary and cumulative effects may be hard to predict or even to understand without an underlying framework of ecosystem relationships.

Traditional forest management techniques have therefore been based on a fairly limited information base, and have often underestimated long term ecosystem impacts while emphasizing resource (timber or wood chip) extraction.

Over more than 30 years, forest ecologists at three universities have studied a single forested watershed in the northeastern United States, trying to understand the detailed ecological processes within a forested ecosystem and thus to improve the knowledge base for forest management. They have attempted to employ a true ecosystem approach in their research, examining all ecological systems as single interacting units, rather than focusing on separate elements of

Figure 11.1: Map of the contiguous United States showing New Hampshire

the complex system. Their main interest in this research has been to draw conclusions about the ability of a forest to recover from impacts such as fire or logging, and thus to make recommendations about preferred logging practices in the northeastern United States.

? 2. What Problem Are We Trying to Solve?

The small watershed approach

The Hubbard Brook researchers—Herb Bormann, Gene Likens, Robert Pierce, John Eaton, and others— adopted a method they call the "small watershed approach" to understand the workings of a complex forested ecosystem. Their reasoning was that small watershed units could be understood more easily and more completely than large, complex watersheds.

Although they studied a series of small watersheds covered in second-growth forest, their conceptual approach began with the hydrologic cycle. In humid regions, like northeastern North American, water acts as a sort of "lifeblood" in the system—a means of transporting and cycling materials. They therefore looked for a terrestrial ecosystem underlain by impermeable bedrock, so that the only inputs to the system would be meteorologic and biologic. They further assumed that there would be no transfer between watersheds, so geologic inputs could be disregarded: all geologic inputs would eventually turn up in the streams draining the watershed. By measuring the meteorologic input and the geologic output of nutrients, they could then arrive at the net gain or loss of a given nutrient in the ecosystem.

The Hubbard Brook Experimental Forest

Bormann and his colleagues chose for their research an experimental forest tract, the Hubbard Brook Experimental Forest, established in 1955 by the United States Forest Service as the principal research area for the management of watersheds in New England. The Hubbard Brook watershed is located near West Thornton in the White Mountain National Forest of north-central New Hampshire.

The experimental forest covers an area of 3,076 ha and ranges in altitude from 229 to 1,015 m. Bormann and his colleagues established six sub-watershed research areas within the forest, ranging from 12 to 43 ha in area and from 500 to 800 m in altitude. All of these sub-watersheds are steep (average slope of 20 to 30%) and face south, with relatively distinct topographic divides allowing them to be clearly separated in experimental procedures.

An important feature of the Hubbard Brook area is that it is covered with "second-growth" forest—in other words, not the forest that existed before the arrival of European settlers. The forest was logged in 1910-1919—as were most forests in notheastern North America—but has not been disturbed since. There is no evidence of fire in the history of the forest.

In the view of the U.S. Forest Service, tracts of forest like this are a renewable resource, to be harvested periodically to supply our society's voracious demand for wood, wood products, and paper. The Hubbard Brook ecosystem is in many ways typical of upland forest ecosystems in the northeastern United States. It is not a particularly rare or special ecosystem that deserves special protection: the reality is that society's demand for wood requires that forested ecosystems be disturbed from time to time. The problem of interest to us in this case study, and to Bormann and his colleagues in their research, is how to use this renewable forest resource sustainably, fulfilling our own society's needs while protecting the resource for the use of future generations.

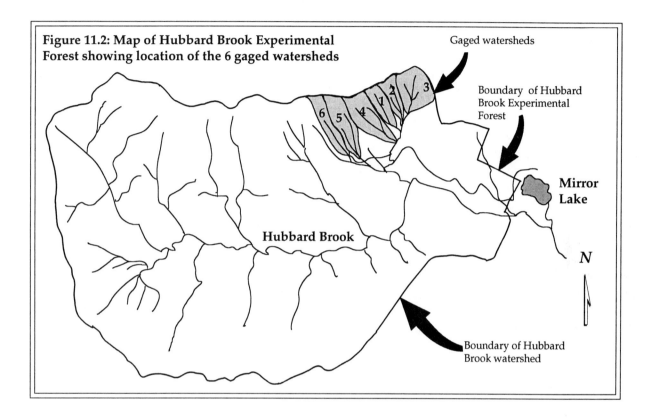

Figure 11.2: Map of Hubbard Brook Experimental Forest showing location of the 6 gaged watersheds

Gaged watersheds

Boundary of Hubbard Brook Experimental Forest

Mirror Lake

Hubbard Brook

Boundary of Hubbard Brook watershed

N

3. *What Components of the Environment Are Affected, and How?*

The Hubbard Brook Experimental Forest is in many ways typical of northeastern forested ecosystems. It does have a few special characteristics that are important in analyzing its function. One of these is its relatively impervious bedrock, which makes it an almost "water-tight" system in terms of the movement of water and dissolved materials. Some other ecosystem characteristics are given in the following paragraphs.

Climate

As a mountain ecosystem, the climate of the Hubbard Brook watersheds varies with altitude. The area is generally classified as humid continental, meaning that it experiences short, cool summers and long, cold winters.

Weather in the watersheds is changeable, partly in response to wind patterns around the White Mountain peaks. There is a large range in daily and annual temperatures, but on the whole, precipitation is distributed equally throughout the year, some falling as rain and some as snow. Prevailing winds are from the west: from the northwest in winter, and from the southwest in summer. Occasionally, cyclonic disturbances bring in easterly winds with moisture carried from the Atlantic Ocean.

The combination of cold winter temperatures and abundant atmospheric moisture contributes to the development of a 1.5-m snowpack in most winters, a layer thick enough that occasional milder temperatures do not fully melt the pack until the arrival of spring. This feature is important in the creation of an insulating layer over the forest soils during winter, allowing even the topmost layer to remain unfrozen during the coldest months.

Geology

Like most other areas in northeastern North America, the Hubbard Brook watersheds were covered with glacial ice during the last (Wisconsinan) glacial period, with soils only exposed about 12,000 to 13,000 years ago when the glacial ice sheet retreated northward.

In the watersheds, the bedrock is highly metamorphosed and deformed mudstones and sandstones of the Littleton formation. These rocks are considered impermeable (rather like a slate roof), thus forming the "water-tight" lower boundary of the ecosystem.

Overlying bedrock is glacial till (mixed deposits resulting from glacial action) that have their origin in the local Littleton Formation rocks. Soils are mostly well-drained sandy loams about 0.5 m deep with a 3- to 15-cm-thick layer of organic matter—humus, including decaying leaves and branches—at the surface. The combination of humus layer and well-drained soils means that most precipitation infiltrates easily into the soil. Overland flow, such as sheet flow or gullying, is rarely seen. Rapid drainage is further encouraged by the rough surface topography—fallen trees, decaying twigs and branches, a variety of herbaceous and woody plant species—and the absence of soil frost.

Flora and fauna

The vegetation in the forest ecosystem is part of the northern hardwood ecosystem, containing a mixture of deciduous and coniferous species that may occur as solely-deciduous or as mixed deciduous-coniferous stands. Common tree species in the watersheds include beech, sugar maple, yellow birch, white ash, basswood, red maple, red oak, white elm, hemlock, red spruce, and white pine. Stands display a range of ages and a wide range of shrubs and herbaceous species; in these characteristics they are entirely typical of a developing northern hardwood forest ecosystem.

Fauna in the Hubbard Brook area are also typical of northern hardwood forests, including more than 90 species of birds, snowshoe hare, beaver, red fox, black bear, and whitetail deer (populations of deer are usually not very dense in this area because of severe winters and high hunting pressure).

Drainage

The experimental watersheds are drained by small, perennial streams (1 to 2% of the watershed area) whose flows range from a trickle during summer droughts to hundreds of cubic meters per hectare per day during snowmelt and storm events. The dense forest cover on the watersheds means that the streams are shaded throughout the summer and fall, with the result that stream temperatures are fairly constant during that period. Stream water is usually saturated with oxygen and exhibits a pH less than 6.

Typically, stream beds in these watersheds, like the forest floor, are physically complex, covered by organic debris, a range of sediment sizes from sand to boulders, and occasional obstruction by fallen branches or other debris. Stream flow therefore passes from ponded area (behind a fallen branch) to a riffle (fast-flowing) area—a classic "pool-and-riffle" configuration.

Precipitation in the watersheds is distributed throughout the year, with about 69% (roughly 90 cm) falling as rain and about 31% (about 40 cm) falling as snow. Given that the watersheds can be considered impervious, and all streams are headwater streams originating within the watershed, 100% of the incoming water can be assumed to come from precipitation. Of this, 63% (about 83.3 ± 6.6 cm/unit area $\pm$ standard deviation) flows out of the system as streamflow, while the remainder (37%, or 48.9 ± 1.0 cm/unit area $\pm$ standard deviation) is evaporated or transpired.

Box 11.1: Productivity in the Hubbard Brook Experimental Forest

Net primary productivity averages 1,040 g/m²-yr dry weight

Living biomass accumulation averages 360 g/m²-yr dry weight

Forest floor accumulation averages 33 g/m²-yr dry weight

Clear-cutting methods

The Hubbard Brook researchers aimed to use their research findings to help make decisions about forest management in the northeastern U.S. To do this, they conducted controlled experiments using a variety of clear-cutting methods, to determine the effects of each on the forest ecosystem. The methods they examined were as follows:

Stem-only harvesting

In this older clear-cutting method, a common one in the White Mountain region, all merchantable trunks between stump and crown are harvested for use in lumber, millwood, and pulpwood.

Whole-tree harvesting

In whole-tree harvesting, the entire tree above ground (but excluding the root system) is cut off and chipped for use in paper pulp or reconstituted paper products. This method has been used only in the last 20 years. The method can be applied in different ways, with different environmental consequences. Generally speaking, however, whole-tree harvesting results in considerably more soil disturbance than stem-only harvesting because it requires heavy machinery to be brought onto the site for tree harvesting and chipping.

Complete-tree harvesting

Complete-tree harvesting is also a newer method, in which a large machine is brought onto the site to remove and chip the entire tree, above ground and below ground. The method essentially involves uprooting the tree and root system, which are then processed for chips. Because larger trees are more difficult to harvest in this way, some new approaches include harvesting of young trees 10 to 20 years in age. Removal of root systems results in increased disturbance of the forest floor and may inhibit regeneration of species that depend on vegetative reproduction or growth from buried seeds.

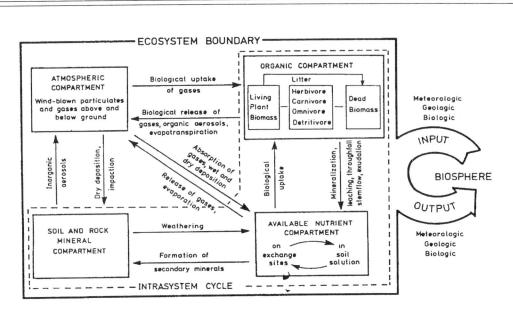

Figure 11.3: A model depicting nutrient relationships in a terrestrial ecosystem. Inputs and outputs to the ecosystem are moved by meteorologic, geologic, and biologic vectors. Reprinted with permission from G. E. Likens, F. H. Bormann, Robert S. Pierce, John S. Eaton, and Noye M. Johnson. 1977. *Biogeochemistry of a Forested Ecosystem.* Springer-Verlag, New York.

Box 11.2: Some observations about maturing forest ecosystems

1. *Ecosystems are open. Water and nutrients continually flux through the boundaries and cycle internally between the various components of the ecosystem. The flow of water, nutrients, and energy is highly regulated by biotic and abiotic components of the ecosystem.*

2. *The stability of the developing ecosystem is characterized by:*

    (a) *An excess of primary productivity over decomposition, allowing biomass to accumulate*

    (b) *Close regulation of the water chemistry and erodibility*

    (c) *An ability to exert strong control over intrasystem aspects of the hydrologic cycle (e.g., water loss by transpiration)*

3. *These processes determine the size of nutrient reservoirs and produce nutrient cycles typified by minimum outputs of dissolved substances and particulate matter and by maximum resistance to erosion.*

4. *The system acts as a filter for atmospheric contaminants, especially H^+, N, S, and P and certain heavy metals, which accumulate within the ecosystem. The undisturbed forest ecosystem therefore acts as a pollution buffer for society.*

Source: Bormann et al. 1977; Borman et al. 1979.

4. How Can I Analyze This Information?

Review published results

Bormann and his colleagues have the advantage of more than 30 years of controlled experimentation in the Hubbard Brook watershed. The average analyst does not have this experience but can take advantage of Bormann's published results to draw conclusions about tree-harvesting practices.

In experimental deforestation (clear-cutting followed by application of herbicides to suppress regrowth) the Hubbard Brook researchers observed the following phenomena:

1. Immediately following cutting, there is a loss of biotic regulation. Hydrologic and biogeochemical parameters, which were observed to be relatively stable and presumably highly regulated in the maturing ecosystem, are grossly altered and for a short time appear to be "out of control".

2. The loss of biotic regulation results in a temporary increase in resource availability, especially for water and nutrients, as well as increased radiant energy at the forest floor. These conditions are ideal for the promotion of rapid regrowth, particularly of opportunistic plants.

3. The first few years following cutting typically show rapid colonization of the site by species whose opportunistic reproductive and growth strategies allow certain of them to exploit the new conditions. Rapid growth results in a gradual return of biotic regulation over ecosystem processes. With time, annual plants such as raspberries are replaced by sunlight-tolerant woody plants such as birch, and over the long-term (as the forest canopy grows) by more shade-tolerant species like maple and beech.

4. It may be many years before water and nutrient cycles return to their pre-cutting levels. Studies in North Carolina have shown that even 23 years after harvesting, deviation from expected streamflow levels was still significant (5 to 10% above normal). As water flows out of the ecosystem, it takes nutrients and other dissolved materials with it.

What are the *mechanisms for these changes?*

Aside from the obvious ecosystem changes associated with removing trees, there are two less obvious impacts of logging on the forest ecosystem. These are road building and disturbance of the forest floor.

Road building

In a typical logging operation, roads cover 2 to 10% of the forested area. Roads are essential to bring in people and vehicles and to allow transportation of the extracted timber or wood chips out of the area. Associated activities can include building of the road itself (grading equipment, gravel trucks, and so on), log skidding, preparation of a seedbed following logging, and similar activities.

Road building cuts into the sponge-like humus layer and topmost soils of the forest floor. Often, excavated material is piled to one side, burying other portions of the forest floor. These measures have three immediate effects. First, they remove the flow-baffling effect formerly provided by the humus layer, allowing water and therefore nutrients to flow more freely. Second, a road cut into the forest floor provides an easy conduit for water, rapidly draining away water that would normally be held in the organic humus layer. Third, rapid drainage can encourage erosion—a common problem on steep mountain roads.

Road building also compacts underlying soils and reduces natural infiltration potential. Flowing water is diverted down the roadway, starving normal flow routes through the insulating humus layer. Snow cover now lies directly on the road surface, permitting freezing and frost heave of the soil surface, and interrupting the gradual seepage that occurs in the undisturbed forest floor.

Disturbance of the forest floor

With road building, introduction of heavy machinery, log skidding, and similar activities, the forest floor usually experiences considerable disturbance in a logging operation. As discussed earlier, the forest floor plays an essential role in the regulation of water and thus nutrient flow in the ecosystem. It is also a reservoir of buried seeds and roots for vegetative reproduction.

Disturbance of the forest floor is therefore the trigger for release of water and nutrients from the ecosystem. Depending on the cutting technique employed, this disturbance may occur to a greater or lesser extent. Extreme disturbance of the floor may prevent regeneration from buried seeds and roots. To overcome this problem, some forestry companies routinely plow and reseed clear-cut lands, arguing that this will encourage rapid regrowth. Bormann and his colleagues argue, by contrast, that further disturbance in this important zone actually reduces the system's ability to recover from stress, by accelerating the loss of water and nutrients and by simplifying physical structures.

Changes in nutrient dynamics

Nutrients are quickly lost following clear-cutting, through several mechanisms. First, there is accelerated drainage from the site due to road building and disturbance of the forest floor. Possibly even more significant, however, is the removal of plant material from the site, whether just the trunk ("stem-only," leaving roots, branches, and leaves on site), "whole-tree" (leaving roots on site), or "complete tree" (removing all parts of the tree from the site). In the undisturbed ecosystem, decaying biomass is an important reservoir of nutrients; when that biomass is removed from the site, the nutrients go with it.

Evidence from Hubbard Brook and other studies has shown that whole-tree harvesting of a 90-year-old stand results in 2.3 times as much nitrogen lost from the system as when stem-only harvesting is employed. Complete-tree harvesting of a 60-year-old stand results in 4.3

times as much nitrogen loss as stem-only harvesting.

Stands that are cut at more frequent intervals, for instance every 30 years (as has been common practice in this part of the world), have not usually been able to regain their precutting nutrient reservoirs or flow regimes. As a result, with each successive cut, the system is starting at a lower flow and nutrient level, thus taking longer to recover from the stress of cutting. After three or four cuts at 30-year intervals, the authors suggest, the system may be unable to regenerate an acceptable "harvest" of timber. The Hubbard Brook results suggest that, in this region, cutting should be done at no less than 65-year intervals, and probably closer to 110- or 120-year intervals. This is dramatically different forest management policy from that currently employed in northeastern North America.

> **Box 11.4: Some conclusions about ecosystem dynamics and logging practices at Hubbard Brook**
>
> 1. *Whole tree or complete tree harvests do not actually increase production, as proposed by some authors, but rather increase harvests at one time while lengthening the time required for the ecosystem to return to pre-cutting nutrient levels. The impact of the cut depends on the method used.*
>
> 2. *Repeated harvests at time periods shorter than needed to regain precutting nutrient levels could cause a general deterioration of the ecosystem, with starting levels lower, and recovery rates slower, after each cut*
>
> 3. *Therefore cuts should be small, conducted in the context of a larger watershed unit, and should be limited to sites with strong recuperative capacity*
>
> 4. *Cutting should avoid sites with steep slopes or thin soils*
>
> 5. *The cutting and harvesting procedure should do minimum damage to the forest floor*
>
> 6. *Roads should occupy the minimum possible area, and buffer strips should be left along stream banks*
>
> 7. *Planned rotation times should be long enough for the system to regain pre-cutting nutrient and organic matter levels; probably at least 65 years in our northern hardwood forests; 110 to 120 years may be more protective*
>
> *Source: Bormann et al. 1977, 1979.*

5. How Can I Use My Findings to Reach a Solution?

Use the decision-making framework described in Chapter 1 to organize your thinking on this problem, as follows:

1. What is the problem?

In Section 2, we identified the problem as being how to use this renewable forest resource sustainably, fulfilling our own society's needs while protecting the resource for the use of future generations.

2. *In what ways do human activities have impact on the natural environment to cause "a problem"? How do these mechanisms give you clues to possible solutions?*

The main human activity we are concerned with here is logging (and the trucking and road-building activities associated with logging). We know from the environmental evidence that we can shorten the interval needed for forest regeneration by leaving as much plant material on site as possible, by minimizing disturbance of the forest floor. Activities that we know disturb the forest floor are road-building and tillage for seedbed preparation.

3. *What governments are responsible for the issue? Whose laws may apply?*

This is an interesting question because the answer will differ from country to country and from place to place within a country. At Hubbard Brook, which is a U.S. Department of Agriculture (Forest Service) site, the federal government would obviously have a major share of the responsibility and the legal "jurisdiction". But state governments and even municipal governments may also have an interest from their own perspectives. It would be wise to explore the question of jurisdiction with Forest Service representatives and other federal and state regulators to clarify this important question.

4. *Who has a stake in the problem? Who should be involved in making decisions?*

In virtually every environmental problem many people and organizations have an interest, sometimes a financial interest, in how the problem is resolved. It's tempting to consider only the obvious "stakeholders"—in this case, the Forest Service and perhaps the logging contractors. In fact, there are likely many other groups who have an interest in the problem, including local residents' associations, environmental non-government organizations, forest industry assocations, and so on. One of the most difficult challenges in environmental problem solving is to decide who should be involved in making decisions. There's no single correct answer to this, although experience has generally shown that, over the long term, implementation is easier and more complete if a wide range of stakeholders has been involved in the decision making process, and if those involved have been given a meaningful role in decision making—not just in information dissemination.

5. *In the view of your decision-making group, what are the attributes of a satisfactory solution? In other words, when will you be satisfied that the problem is "solved"?*

We've established that the problem, for the purposes of this case, is how to use the forest resource sustainably. But what do we mean by this? Later decision making will be much easier if we are clear as to what targets we are striving for. We might, for instance, say that "sustainability" means being able to extract the same mass of wood tissue from the site each time it is logged, regardless of the interval between cuts. Or we might say that "sustainability" means that we will wait until streamflows and water quality have returned to pre-logging levels. Or we might say that our target is to be able to log every 60 years, extracting the same mass of wood each time, and that the conditions in the watershed just before logging will be streamflows of X cubic feet per second; total phosphorus values of Y mg/L; and total nitrogen of Z mg/L. *It doesn't really matter what targets we choose as long as the targets are clear and specific and the stakeholders agree among themselves.* Decision makers may be more concerned about certain criteria (often cost) than others; criteria weights are therefore usually a local decision, based on local concerns.

6.   How will you evaluate (test, compare) potential solutions?

Targets relating to streamflows, water quality and forest biomass are often central in cases like Hubbard Brook. We could use computer simulation tools to predict what will happen under different management scenarios (e.g. cut every 30 years, every 45 years, every 60 years). Statistical analyses can then help us determine whether there is a significant difference between one approach and another. Cost-benefit analysis and environmental impact assessment are other techniques which could be applied to compare solutions objectively.

7.   What are all the feasible solutions to the problem?

A number of different cutting techniques are described in Section 4 and Box 11.3. There are probably many more described in the forest management literature, or you could develop your own proposals based on your understanding of the dynamics of the forest ecosystem.

8.   Which solutions work "best" in terms of the attributes you identified in (5)?

Forest ecosystems are complex, so it's difficult to draw quantitative conclusions about the performance of various management strategies without computer simulation. You could, however, work solely on the basis of costs, insofar as you can estimate those costs, or perform a qualitative environmental impact analysis (see also Cases 15 and 16) to decide which approach is "best".

9.   Which solution will be easiest to implement?

Logging requires equipment and people to operate that equipment, so an important implementation consideration will be the acceptability *to forest industry workers* of the measures you identify as "best". Those that are too difficult or time-consuming to use will quickly be abandoned. Those that are too costly will simply be rejected by the logging companies—or their customers.

10.   What steps are needed for successful implementation? Who will pay? Who will monitor progress?

Because forest management is a long-term proposition, perhaps extending over periods longer than a century, it is important to include detailed information about the timing of cuts in your implementation plan. You may wish to incorporate periodic reviews of available technologies over the implementation period, so that you can take advantage of emerging technologies that are less invasive and destructive. Other important concerns to include in an implementation plan are responsibility for cutting and for monitoring that cutting, and ongoing post-cutting monitoring of forest productivity, stream water quality, and streamflows.

5. *Where Can I Learn More About the Ecosystem, People, and Culture of the White Mountain Region?*

Dozens of articles and a number of books have been written about the Hubbard Brook project over its 30-year history. The following are suggested entries into that literature.

F. H. Bormann and S. R. Kellert. 1991. *Ecology, Economics, Ethics: The Broken Circle.* Yale University Press, New Haven, Connecticut.

F. H. Bormann and G. E. Likens. 1979. *Pattern and Process in a Forested Ecosystem: Disturbance, Development and the Steady State Based on the Hubbard Brook Ecosystem Study.* Springer-Verlag, New York.

C. A. Federer. 1973. Annual cycles of soil and water temperatures at Hubbard Brook. USDA Forest Service Research Note NE-16. (Contribution 56 of the Hubbard Brook Ecosystem Study). Forest Service, U.S. Department of Agriculture, Upper Darby, Pennsylvania.

G. E. Likens, F. H. Bormann, Robert S. Pierce, John S. Eaton, and Noye M. Johnson. 1977. *Biogeochemistry of a Forested Ecosystem.* Springer-Verlag, New York.

Hubbard Brook data is also available on-line at *gopher://hbrook.unh.edu* or through "The Source of the Brook," a public access service for Hubbard Brook data and information. SOTB is a PC-compatible bulletin board system that can be reached at :

Telephone Number: 603-868-1006
Available: 24 hours a day except during maintenance
Baud Rate: 300, 1200, or 2400
Parity etc.:   8-bit words, no parity, 1 stop bit
Line terminator: CRLF (carriage return-linefeed)

The Hubbard Brook data manager, Cindy Veen, can be reached in person at 603-868-5692.

Botswana

"How can we supply reliable drinking water to an area with recurring drought?"

1. What Is the Background?

Drinking water shortages

Developed countries with abundant freshwater resources tend to be wasteful water users. Canada and the United States, for example, use about 3 times as much water as most European countries and about 10 times as much as water-poor less-developed countries. For many people in the world, obtaining water is a major life-or-death activity.

Botswana, in the south of Africa, is a case in point. Southwestern Botswana's annual rainfall averages only 250 mm, compared to about 1,200 mm in New York City and Vancouver, Canada.

Botswana, the Kalahari Desert, and remote indigenous peoples

About 80% of the country of Botswana is covered by Kalahari sands (see Figure 12.1). The desert is "drained" by a system of dry rivers that seldom carry water even in years of abundant rainfall; shallow depressions or "pans" may collect water temporarily before it infiltrates into deeper groundwater supplies. Groundwater flows are often located 100 m or more below the land surface, and in these dry areas existing flows are often meager and salty. Yet as much as 80% of the local population may rely on these aquifers for drinking and cooking water.

The Kalahari is home to a number of indigenous peoples, including those formerly known as bushmen (the San or Basarwa peoples) and Hotentots (the Bakgothu people). Few of these people now retain their traditional hunter-gatherer lifestyles, most having been settled in permanent or semipermanent settlements under Botswana's Remote Area Development Programme, begun in 1977. With the rise of mining and cattle farming in the area, more and more boreholes are drilled into scanty water supplies, and traditional water sources such as sip-wells, melons, and wild tubers are gradually disappearing. As a result, desert people have begun to settle

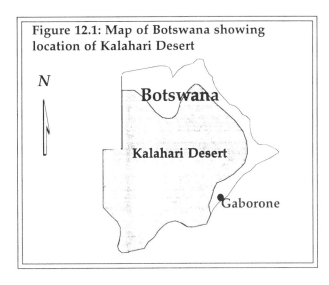

N

Botswana

Kalahari Desert

Gaborone

around boreholes and to guard their water supplies jealously. In the largest villages, water consumption has remained steady at about 15 L/person/day, but for people with private connections to water sources, consumption may be as high as 80 L/person/day, with water used not only for cooking and drinking but for washing and gardening. In the most remote areas, the government has set a target consumption rate of 2 L/person/day. (This is in stark contrast to consumption rates in the U.S. and Canada, which can range well over 300 L/person/day.)

Boreholes are no longer a reliable water source

In recent years, rainfall in the Kalahari has been less than expected, and groundwater recharge has almost stopped. The boreholes that once provided adequate drinking water supplies for small remote populations are now often too salty to drink. The future of settlements that depend on these boreholes is uncertain unless a reliable water supply can be found. Some now receive drinking water brought in by trucks from desalination facilities, but the government is unenthusiastic about continuing this costly and cumbersome approach.

2. *What Problem Are We Trying to Solve?*

Clarifying the goals

The problem here seems pretty clear: to provide a reliable, high-quality drinking water source for remote areas with chronic drought. To identify a "best" solution, however, we should probably be much more specific about the parameters of this decision. For example, how many people should our solution realistically serve? A solution that serves 20 people on a daily basis may not be adequate if we have 800 people to supply.

Similarly, how much water should each person be allowed to take from the communal source? The facts from Botswana show a range from "basic need" (2 L/person/day) to "luxury" (80 or more L/person/day—enough for cooking, drinking, washing, and even garden watering). To decide among possible solutions, we need to be precise about the yield of potable water that we will consider "adequate."

Finally, we need to be precise about the quality of water we want. If current supplies are "too salty", how much less salty do they have to be to be acceptable? What about other considerations such as suspended solids, bacteria, and similar constituents? Some of these, like bacteria, may be very important in our decision, while others, like suspended solids, may be of more concern for aesthetic reasons than for human health.

It is very important to have this "wish list"—a clear and preferably quantitative idea of the problem and a target solution—clear before proceeding to choose among possible solutions. Without this clear target in mind, it's easy to get confused by competing technologies, which may be quite different in cost, appearance, and acceptability to the local population. As an illustration of this confusion, imagine the problem of trying to come up with a "best" technology for conveying information to school children. We're familiar with books as one possibility, but what about television, radio, lecturers...Come to think of it, what about dance, song, theater...? And so on. Until we know what we mean by "best" we have difficulty comparing disparate choices.

Constraints and criteria

Constraints and criteria offer a useful way of sorting out "needs" versus "wants" and thereby facilitating an orderly screening of potential solutions.

Constraints are the limits beyond which you may not go. For example, a major constraint to water supply in Botswana is cost. Drilling wells in the desert is very costly—about $35 to $50 U.S. per meter of depth, with wells often drilled more than 250 m before the search for water is abandoned and another site is sought. Adding the physical support (casing) and protection for the wellhead adds additional costs. A second important constraint is the lack of skilled labor available locally. In developing constraints for this analysis, it would make sense

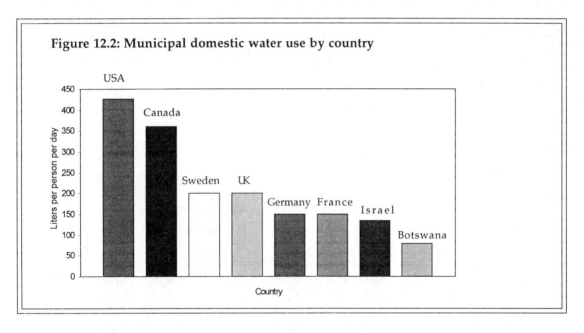

Figure 12.2: Municipal domestic water use by country

to include a dollar limit on expenditure, a limit to the amount of skilled labor needed to implement the technology, and perhaps a limit on the time needed for implementation. We might also want to set a limit on the yield of the technology, expressed in liters per day, and a limit on the number of people the technology is expected to serve. Finally, we would want a technology that supplies water continuously, not sporadically, so this also would be a constraint on the solution.

Constraints are useful in the initial stages of assessment because they allow the analyst to take a long list of suggested solutions and screen out those that are not feasible for the situation of interest. For Botswana, it would be unlikely that a technology that yielded fewer than 10 L/day would be a feasible option, unless it was so cheap that each person could use it. Similarly, a technology that works only when it rains would not be adequate for our purposes, unless water could be stored for supply through dry weather periods.

> **Constraints:** *The fixed limits to your decision, such as the maximum cost or size.*
>
> **Criteria:** *The standards or measures by which something is judged.*

Criteria are the factors you use to choose among a set of feasible options—to decide which is truly "best" for your purposes. If we have several feasible options, we might for example choose the cheapest as "best" if cost were a criterion in our analysis. Or if several technologies supply adequate drinking water quality, but only one supplies water of excellent quality (expressed in terms of bacteria, solids, or some other measure), then we might choose that technology on the basis of its performance.

Choosing good criteria is tricky. Often, it makes sense to discuss your suggested criteria with other stakeholders in the problem. In this case, it might make sense to discuss them with local villagers and representatives of the Botswana government. The more people that scrutinize your analysis, the better able you will be to capture the forces that are genuinely important in the local situation.

 # 3. *What Components of the Environment Are Affected, and How?*

The biophysical environment

Climate

The area involved in this issue is the arid desert region of Botswana. In this area, rainfall averages less than 450 mm annually and daytime temperatures are high for much of the year. In recent years, rainfall has been less than expected, forcing many people to rely on water from boreholes. Groundwater recharge has been reduced in recent years, because of lower rainfall, so the yield from boreholes is now often scanty and saline. In some areas, more than 50% of the boreholes are salty or dry. Even boreholes that at one time yielded sweet water may turn salty or become exhausted because of increasing use and diminished recharge.

Water supply

About 20% of Botswana's population of 1.5 million live in the desert, many clustered around the boreholes. When desert dwellers are unable to secure adequate quantities of drinking water from boreholes, they may obtain it from trucked supplies, from reticulation (that is, piping the water from a more distant sweet borehole), or from desalination using a variety of techniques. Roads through remote areas are sandy and difficult to drive on. Cars and trucks must often limit their speed to 15 km/hr or less. Yet trucking is the only available method to obtain potable water for 70 or so settlements in Botswana.

Available materials

In many less developed countries, including Botswana, it can be difficult to obtain high-quality construction materials, or even basic materials like sand. The local population may not have expertise in construction techniques. These factors can be very important not only in the initial building phases of a project but also—perhaps even more critically—in the operation and maintenance phases. Where a structure has been built with poor materials, it may deteriorate more quickly than we would expect in a more developed country. Faster deterioration of course means earlier and possibly more frequent repair, hence the significance of ongoing availability of high-quality materials and skills. Training programs can help alleviate skills deficiencies, but again they must be comprehensive and ongoing to be effective.

In Botswana, there is an existing construction industry that employs materials from South Africa. Fiberglass and brick are both produced within Botswana. Concrete is more problematic, since the local sand (which provides an essential matrix for concrete) is very fine and uniform and thus reduces the strength of locally produced concrete.

Available technologies for water supply

There are three main methods available for the provision of potable water in arid regions like Botswana. Each method has positive and negative aspects.

Trucking

Trucking is currently the only method available for many remote settlements, even though it is expensive and unreliable. Who gets water, when, and where may be dictated by social factors (for example a wish to visit a particular village) rather than strictly by need.

Piping

Conveying water from boreholes via pipelines is another possibility where adequate borehole supplies exist. This method does require pumping, however, so it is limited to shorter distances, usually 40 km or less from the nearest borehole.

Desalination

Desalination is a newer technique, or rather a group of techniques, that can provide potable water where there is a continuous supply of saline water.

How do we choose which of these methods is "best" for rural Botswana? One approach is to use multicriteria evaluation to assess the performance of each water supply method on the various decision criteria. This analytical approach is discussed in more detail in the following sections.

 # 4. How Can I Analyze This Information?

Determining the impact of different technologies

In an earlier section, we discussed the process of establishing constraints and criteria for an "acceptable" technology. We now have a reasonable understanding of the environment within which the chosen technology is to operate. How can we determine which technology is "best" for the environmental system of interest?

One way to approach this problem is through multicriteria evaluation, using constraints and criteria to guide the process. This is normally done in two stages.

Applying the constraints

Once possible technologies have been identified, a next step is to decide which of these technologies are even feasible, or practical, for the area in which we are working. We do this by determining which technologies meet the constraints we have imposed on our decision. Remember that constraints place limits on your decision—limits of space, cost, performance, skill level, and so on. There is no rule of thumb for developing constraints. Normally, they are established in consultation with the stakeholders in the problem: the users of the technology, the regulators, the manufacturers, and others who have a clear interest in the outcome of the process.

As an example, let us say that we have placed a limit of $200 a year in capital (purchase or construction) costs, $50 a year on maintenance, and basic literacy as the only skill required by an operator. We can then examine each option for the provision of drinking water and determine which meets our basic requirements. If we learn that the cost of operating a truck greatly exceeds $50 a year, we should eliminate that option from further consideration (unless we can modify it—perhaps by using a smaller truck—so that it *does* meet our requirements).

Table 12.1: Example application of constraints

	Capital cost (annualized)	Operation and maintenance	Required skill level
Option 1: Trucking	$2,000	$100	Literacy (assumes user is not driver)
Option 2: Piping	——Not feasible—distance is too great——		
Option 3: Desalination	——Depends on method——		

When all options have been evaluated against the project constraints, you should be left with a shorter list of feasible options. To find out which of this list is the "best" choice, given the priorities we have set for our analysis, we must evaluate each option against our decision criteria.

Application of constraints may also provide valuable insight into the analysis. For example, Table 12.1 shows that an important feasibility criterion, ability to function even when distant from freshwater sources, has been omitted from the preliminary analysis. In this case, it would be wise for the decision-makers to amend their constraints to capture this very important consideration.

A second insight that can be gained from application of constraints is that some options are not in fact single choices but rather

Table 12.2: Example application of constraints (revised)

	Capital cost (annualized)	Operation and maintenance	Required skill level
Option 1: Trucking	$2,000	$100	Literacy (assumes user is not driver)
Option 2: Piping	——Not feasible—distance is too great——		
Option 3: Desalination			
3a: Shallow-basin	$50	$10	Strength
3b: Reverse osmosis	$100	$35	Literacy
3c: Solar distillation	$100	$50	Literacy

groups of choices. Using Table 12.2 as an example, the option of "desalination" is not easy to assess against the project's constraints simply because different desalination methods differ so widely. In this case, the analyst could develop several suboptions to capture the differences: option 3a could be shallow-basin solar stills; option 3b could be reverse osmosis; option 3c could be solar distillation, and so on. (These technologies are described in greater detail below.)

Applying the criteria

Within a group of feasible options, we still have the problem of choosing which option is "best". We do this using certain measures that are important to us in the particular problem. This is exactly the same problem as choosing which new car to buy. If we set the maximum cost of the car at $20,000, a number of options are automatically excluded (did not meet our cost constraint), while several others remain. Of these, we may make our choice depending on each model's seating capacity, gas usage, horsepower, or cost, or some combination of these. It is interesting to note that some factors, such as cost, may be both constraints (in that they have maximum allowable values) and criteria (in that the least cost option may be the most desirable).

It is unrealistic to expect that the same set of constraints and criteria will work for every situation or every group of stakeholders. After all, you wouldn't want other people choosing your new car for you—they may not base their decision on the same criteria you would. It is therefore important that all stakeholders agree on which constraints and criteria are appropriate and also on how each should be measured.

Box 12.1 shows one way in which criteria could be applied to our water supply problem. Clearly, the criteria shown in the example are at best subjective and at worst arbitrary. Nevertheless, they provide a mechanism by which you can reflect in your decision-making the consensus of the stakeholders as to which factors are important. Sometimes, weights are applied to the criteria to reflect relative importance. Althought it is not essential, it can be helpful if any such weights sum to 100% to avoid grossly overweighting one or more factors.

The example shows only qualitative "scores." It is certainly possible to replace "high-moderate-low" scores with numerical values, such as yield expressed in L/day or costs in dollars. It is also possible to mix qualitative and quantitative criteria in a single table.

Analyzing a multiattribute decision matrix

There are probably as many ways to analyze a decision matrix as there are ways to compile one. You can read about what other analysts have done by examining the literature, but you may find it equally useful to develop your own approach. The most important thing in such an analysis is to evaluate all options on a consistent basis. A few simple suggestions may help:

If you can reduce all scores to a common measure

If you can reduce all your scores to a common measure (dollars are a good example), you can simply add the scores for each option (weighted as appropriate) to arrive at a total value for that option. The option that minimizes costs would then be chosen as "best". (In some cases, you may choose to apply a ranking or other numerical score; the choice of a "best" option would then depend on how you have structured your ranking system.)

If all scores are qualitative or a mixture of qualitative and quantitative

If, as in our example, your scores are all qualitative, or you are using a mixture of types of scores, you can do several different things:

Box 12.1: An example

Let's say that we have decided that some form of desalination is the only feasible method for sustainable drinking water supply. We may have justified the elimination of trucking because of its high cost (failed to meet our cost constraint) and piped distribution because it doesn't perform well when water must be distributed at great distances from a sweet water source.

Which method of desalination is "best" for rural Botswana? To make this decision we must establish a common basis of comparison: the decision criteria. What criteria should we use? Although there is no easy answer to this without on-site investigations and consultation with stakeholders, we can propose some factors that we think are important in choosing an appropriate technology.

*Examples of criteria that might be important to us are **cost (again), failure rate, availability of parts for repairs, sensitivity to the quality of raw water, and electric power demand.** We can use these criteria to evaluate several possible desalination technologies in a "decision matrix," as illustrated below.*

Still Type	Description	Cost	Parts	Failure Rate	Sensitivity	Power	Yield
Shallow basin	Shallow pit covered with tilted sheet glass	Low	Good	Low	Low	None	Low
Reverse osmosis	Salty water forced against a membrane that allows water but not salt to pass	Moderate	Poor	Moderate to high	High	Moderate	Moderate to high
Ghanzi still	Copper coil immersed in cool salty water; water boiled in separate container and steam piped through coil to condense	Low	Good	Moderate to high (if poorly maintained)	Low	Needs large amounts of fuelwood	Low to moderate
Fiberglass solar still ("Mexican still")	Molded basin covered with tilted sheet glass; stair-step design improves efficiency	Moderate	Good	Low	Low	None	Moderate

1.   You can look for an option that "dominates" the analysis, that is, is "best" in every category. That option is then chosen as "best" with no further analysis.

2.   You can look for the option that does best under worst case conditions. For example, if political or economic circumstances lead you to believe that availability of parts could be a significant problem in the future, choose the option that has the highest yield yet requires only locally available materials.

3.   You can choose the option that does "best" on the factor(s) that stakeholders consider most important. For Botswana, these might include failure rate, sensitivity to influent water quality, and yield.

Communicating Your Results

It is to be hoped that your data collection, development of contraints and criteria, and decision analysis will have included input from local stakeholders in the problem. These stakeholders serve two important functions in your analysis. First, they bring to the analysis perspectives, values, and information that may be outside your own experience but centrally important to the decision at hand. For example, they may know from experience how easy or difficult it is to obtain high-quality fiberglass in Botswana, how much it costs, how well it stands up to local weather conditions, and so on.

The second reason to include local stakeholders is to provide an essential link back to implementation. An environmental decision is of no value if it will never be used. If local stakeholders have been involved in the decision in a meaningful way—in other words, not just informed about the "experts'" decision—they can be the conduit through which your decision is communicated to those affected by it.

Stakeholder involvement and communication strategies are discussed in more detail in Case Studies 4 and 7.

5. How Can I Use My Findings to Reach a Solution?

Selection of a "best" drinking water supply for Botswana is a fairly straightforward matter of developing constraints and criteria, then evaluating alternative technologies against them.

1. What is the problem?

In Section 2, we identified the problem as choosing a "best" method of providing a safe and reliable drinking water supply for rural communities experiencing chronic drought.

2. In what ways do human activities have impact on the natural environment to cause "a problem"? How do these mechanisms give you clues to possible solutions?

This case describes an ecosystem that is only marginally viable for human habitation, especially under the drought conditions of recent years. Human impact on this environment is in fact minimal and relates mainly to drinking water extractions. In analyzing this case, we are in a sense concerned with *environmental* impacts on *humans* through drinking water scarcity. The scattered settlements and poor infrastructure do, however, suggest that we should be seeking solutions that do not require much transport, construction or maintenance to be successful.

3. What governments are responsible for the issue? Whose laws may apply?

The Botswana federal government has a responsibility to oversee the public health of citizens of Botswana, and in this regard provision of safe drinking water could come under their jurisdiction. International aid and development agencies could conceivably have an interest in the matter and may or may not be affiliated with the governments of other countries. Local (village) governments would clearly be concerned with access to a reliable water source.

Who has a stake in the problem? Who should be involved in making decisions?

Stakeholders in this case potentially could include a wide range of players, including the governments of Botswana and other countries, local governments, international aid and development agencies, manufacturers and suppliers of pipes, pumps, and desalination technologies, local construction companies, and of course the consumers of the water supply.

5. *In the view of your decision-making group, what are the attributes of a satisfactory solution? In other words, when will you be satisfied that the problem is "solved"?*

Section 4 discusses in detail the development of constraints and criteria for a multiattribute decision-making exercise. This section gives a number of examples of decision criteria; others are certainly possible, depending on the interests of the decision-making group.

6. *How will you evaluate (test, compare) potential solutions?*

Section 4 discusses multiattribute decision analysis in detail. This is a very useful method for evaluating alternative technologies like those for desalination. Other approaches such as pairwise comparison are described in Case Study 6 on India.

7. *What are all the feasible solutions to the problem?*

There are several main methods of providing drinking water to rural Botswana, as described in Section 3. Within each category, a number of specific technologies may be available. Some of the available desalination technologies are described in Box 12.1; many others are probably to be found in the literature or even in equipment suppliers' catalogs. You may also wish to design your own desalination equipment using some of these technologies as a starting point.

8. *Which solutions work "best" in terms of the attributes you identified in (5)?*

This step follows readily from the multiattribute decision matrix analysis described in Section 4. Be careful weighting decision criteria: it's best to make sure that all weights add to 100 (100%) to avoid grossly exaggerating the performance of any one option. You need not use weights at all, however—you may simply decide that all your decision criteria are of equal importance.

9. *Which solution will be easiest to implement?*

Ease of implementation in this case will likely relate to maintenance requirements and the availability of spare parts, lubricants, and so on. Technologies that are difficult to operate or which have a high repair frequency may be unattractive to many in remote communities. Simpler, less expensive technologies are likely to be more successful even when their yield is less than that from a high-tech pumping/piping system or an automated still.

10. *What steps are needed for successful implementation? Who will pay? Who will monitor progress?*

Several levels need to be considered in planning implementation of drinking water supplies. First, the providers and funders of the technology must be identified and their responsibilities clearly laid out. Local transportation arrangements may be complex in remote communities and should therefore also be planned in detail. Successful implementation may also require the development of local "capacity", as it is often called in the literature on international development. "Capacity" means the ability of the local community to understand, operate, and maintain the technology. Building capacity may require formal education programs, periodic visits from technical support experts until the technology is running smoothly, and similar considerations.

 # 5. *Where Can I Learn More About the Ecosystem, People, and Culture of Rural Botswana?*

The following sources are useful background on Botswana's people and culture and the problems of desalination in remote arid locations.

R. Matz. 1965. Desalination of sea and brackish water: the present state of the art in Israel. In: A. Girelli (ed.), *Fresh Water From the Sea*. Proceedings of the International Symposium held in Milan by Federazione Delle Associazioni Scientifische e Tecniche and Ente Autonomo Fiera di Milano. Pergamon Press, Oxford, U.K.

D. E. Osborn, R. Sierka, and M. Latif. 1988. Water problems, solar solutions: applications of solar thermal energy to water technologies. In: J. R. Starr and D. C. Stoll (eds.),The Politics of Scarcity: Water in the Middle East. Center for Strategic and International Studies. Westview Special Studies on the Middle East. Westview Press Inc., Boulder, Colorado.

J. R. Starr and D. C. Stoll (eds.), 1988. The Politics of Scarcity: Water in the Middle East. Center for Strategic and International Studies. Westview Special Studies on the Middle East. Westview Press Inc., Boulder, Colorado.

M. A. Ukayli and T. Husain. 1988. Comparative evaluation of surface water availability, wastewater reuse and desalination in Saudi Arabia. *Water International* 13(1988): 218-225.

R. Yates, T. Woto, and J. T. Tlhage. 1990. Solar-Powered Desalination. International Development Research Centre, Ottawa, Canada.

Colorado

"How can we prevent pollution from a manufacturing operation?"

1. What Is the Background?

Often, pollution means wasted raw materials and lost revenues

Almost every manufacturing process involves some waste, whether it is solid waste such as packaging, liquid waste such as process effluents, or smokestack emissions. Traditionally, our society has assumed that pollution is the inevitable consequence of economic development—that it's not possible to have economic prosperity without paying the price of environmental degradation.

Over the past 20 years, some industries have begun to challenge this way of thinking. Minnesota's 3M corporation has been a leader in this. For more than 25 years they have had a corporate "3P" policy: pollution prevention pays. What they mean by this motto is that wastes discharged down drains and up chimneys contain valuable resources. Raw materials carelessly spilled then washed away can represent a significant cost to the manufacturer. Wasteful water use can also represent a high financial cost, especially in areas where water is scarce and water prices are high. 3M has consistently maintained—and demonstrated over many years of public reporting—that where a company can *prevent* the creation of wastes, it can actually save money in the long term, even when pollution prevention requires expensive changes in equipment or processes.

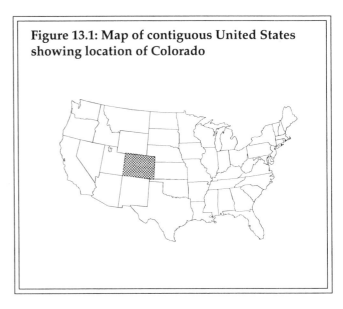

Figure 13.1: Map of contiguous United States showing location of Colorado

Pollution prevention, not just "control"

Much of North America now has "pollution prevention" legislation in place, encouraging industries to take a hard look at their processes and find out where they can eliminate pollution before it occurs. Although costly equipment changes are sometimes necessary, much more often simple "housekeeping" measures, such as careful handling of raw materials, spills and leak prevention, and water reuse, can achieve tremendous results.

With this good progress behind them, some companies are beginning to explore new approaches to pollution prevention. One of these is called "design for the environment" or "green design". Green design implies that environmental protection is built into the product. Sometimes this means that a product is designed for ultimate disassembly so that its components can be reused in the same or a different application. In other cases, processes are completely redesigned to use raw materials and feedstocks that are less toxic or more recyclable.

Coors: A corporation emphasizing "green design"

As the costs of waste disposal escalate, more and more companies are turning to this preventive approach. Among them is Coors, one of America's best-known breweries. Like many large corporations, Coors has spent hundreds of millions of dollars on end-of-pipe pollution abatement, including wastewater treatment, solid waste disposal, and air pollution control. In 1992 alone, the company spent $23 million U.S. on pollution control. Unlike many other companies, Coors has also spent more than 20 years trying to eliminate waste, not just dispose of it after the fact. In part, this arises from Coors' corporate philosophy, which emphasizes technology, ingenuity, and self-reliance.

Coors has found that careful waste audits often prompt ideas for new processes, and that these ideas can somtimes spin off into subsidiary companies that are themselves profitable. In this way, pollution prevention at Coors has meant not only reduced resource consumption and waste disposal costs, but also increased profitability through new ventures.

Coors has faced a number of challenges in adopting this preventive approach. Among them has been the fact that our legal structures tend to reflect our society's view that pollution is inevitable: they stress strict compliance with rigid standards set by a regulatory agency. When a company is experimenting with new processes and technologies—or even with better housekeeping—it can be hard to meet the letter of the law one hundred percent of the time. Management structures may also have to change to encourage flexibility, creativity, and recognition of worker insights.

2. What Problem Are We Trying to Solve?

Economics versus environment?

The problem faced by Coors, as by most other manufacturing facilities, arises from the combination of economic forces (high costs of raw materials and waste disposal) and tightening legal requirements for pollution control. The problem is not simply how to reduce waste, but how to make that reduction effectively and efficiently without creating other pollution problems or resulting in a net cost to the company.

Most environmental managers now espouse the "ecosystem approach" to environmental problem-solving. By this, they mean that it's not enough to look at a single aspect of the environment—for instance, the water—in solving a problem . It's necessary to examine all the interconnected components of the environment as a functioning system, both to understand the system as it currently works and to anticipate the impacts on the system of change. "Air stripping" benzene from a liquid effluent stream (that is, exposing the benzene-rich stream to excess air to encourage volatilization) may solve the water problem and allow the company to meet water regulations, but it may create an air pollution problem and put the company out of compliance with its air pollution limits. Unless we study all the aspects of the system, we won't detect these interrelationships.

> *"All pollution and all waste is lost profit."*
>
> *Bill Coors, Chairman*
> *Coors Brewing Comapny*

An industrial "ecosystem"?

Can we really think of a built environment like a factory as an "ecosystem"? Sure! In fact, there's not much difference between the way we would analyze a natural ecosystem like Hubbard Brook (Case Study 11) and the approach we would take with a constructed system. In describing either, we would want to come to an understanding of several key factors:

What materials are of most interest to us in the system?

In Hubbard Brook (Case Study 11), the research scientists were most concerned with plant nutrients such as nitrogen and with the building blocks of photosynthesis. In an industrial setting, we might be concerned about volume of solid waste, or a chemical of particular human health or environmental concern, like benzene. We could also examine things like energy flow, labor demand, or water use; similarly, we could choose a combination of these "indicators" of most interest to our analysis. The word "indicator" is sometimes used to suggest a measurement that can be made to assess progress against targets—a sort of indicator of the state or health of the overall system.

What are the main reservoirs of these materials within the system?

Once we have identified indicators that are of particular interest to us, we can examine the system in detail to identify important reservoirs. If we are interested in plastic packaging, we might identify the raw materials storage area as one "reservoir" and temporary waste storage containers as another. If we are interested in a particular chemical, we can measure the concentration of that chemical in air, water, and solids at every step of the industrial process to determine the steps where most of that chemical is held or generated.

What processes move the materials from one reservoir to another?

The Hubbard Brook project was largely concerned with defining ecosystem processes, and the same should be true in examining an industrial ecosystem. What goes on at each manufacturing step? These steps are often referred to as "unit processes" because each can be isolated for engineering, chemical, or other analysis. Are there certain unit processes that can be eliminated from further consideration, simply because they do not contribute to the generation, storage, or movement of the indicator of concern?

How fast does this movement occur, and what factors govern these rates?

The Hubbard Brook scientists were also very concerned with the *rates* at which processes occur. They looked at the rate of conversion of ammonia to nitrite and nitrate, for example, and the rate of water loss from leaf surfaces through transpiration. We can do the same in the industrial ecosystem. How fast does waste plastic move from the raw materials storage area to the temporary waste storage container? This rate is probably related to the rate of materials usage, the form in which raw materials are bought and/or received, and similar factors. Or, how fast is benzene used in a given unit process? At what rate is it released to the surrounding air? And so on.

When we understand these factors, we can begin to understand the system as a functioning entity and thus begin to anticipate the impact of changes we might as managers impose on the system. In the context of industrial pollution prevention, the problem therefore has three parts:

1.   To gain an understanding of the industrial ecosystem under consideration, with respect to the creation, storage, and movement of key materials of interest

2.   To identify the most important sources of waste, for instance in terms of proportion of total volume generated from the facility

3.   To select waste reduction options that will effectively reduce the waste while maintaining a net profit for the operation

In terms of selecting remedial options, the U.S. (federal) Pollution Prevention Act sets out a very clear hierarchy of pollution prevention approaches. Similar hierarchies are described throughout the pollution prevention literature. Coors has elaborated on the act's hierarchy for its own special applications, in particular its brewery and container operations. The following figure illustrates Coors' expanded pollution prevention hierarchy showing Coors' program options.

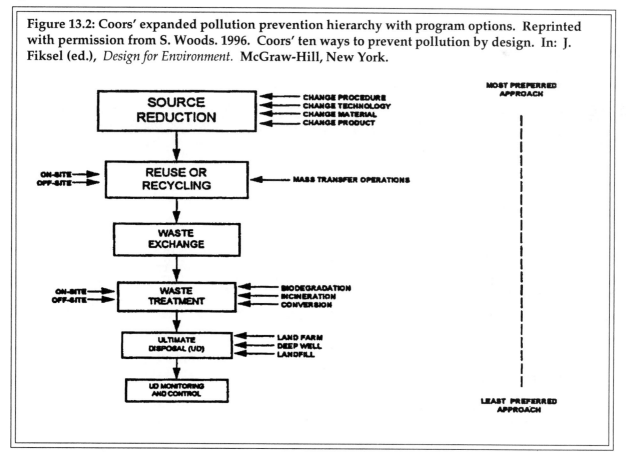

Figure 13.2: Coors' expanded pollution prevention hierarchy with program options. Reprinted with permission from S. Woods. 1996. Coors' ten ways to prevent pollution by design. In: J. Fiksel (ed.), *Design for Environment.* McGraw-Hill, New York.

From this heirarchy, Coors has developed a series of source reduction options which they have summarized in Figure 13.3.

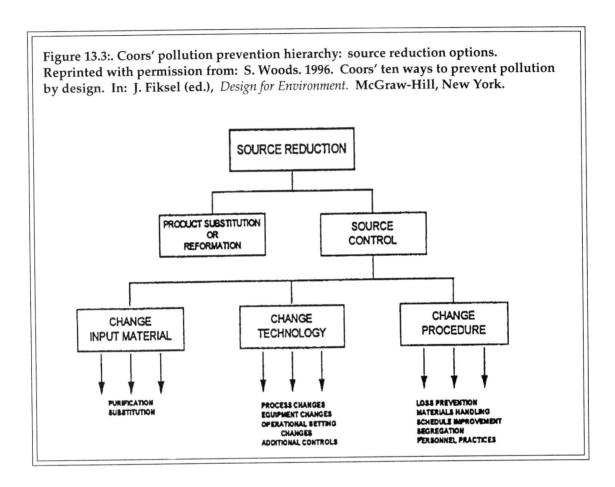

Figure 13.3:. Coors' pollution prevention hierarchy: source reduction options.
Reprinted with permission from: S. Woods. 1996. Coors' ten ways to prevent pollution by design. In: J. Fiksel (ed.), *Design for Environment.* McGraw-Hill, New York.

A tool and a risk?

Assessment of the overall industrial ecosystem is a highly effective way to understand the sources and movement of wastes and to develop sound reduction methods. It can also be risky for the industry, however. Detailed audits can turn up process weaknesses or pollution sources that had been overlooked by regulatory agencies. Reporting these to regulators may seem ethical and environmentally sound, but it can carry the risk of closer scrutiny in future, and even punitive fines. In 1991, Coors undertook an extremely detailed inventory of emissions from its brewery operations, in part to extend and validate assumptions made by the U.S. Environmental Protection Agency. Coors found that EPA's estimates of volatile organic compound emissions were in fact too low, and concluded that brewery operations were a significant source of VOCs.

The company was not required by law to spend the $1 million or so it cost to undertake this study, and they were not required to report their findings to the government. In doing so, they incurred a $237,000 fine from the state government and an enforcement action that lasted more than a year. This kind of reaction from government is to be expected in a traditional command-and-control system that emphasizes end-of-pipe pollution clean-up, not prevention. It's not, however, a satisfactory response to a company's efforts to go "beyond compliance". Government structures and policies may have to change to encourage companies to go beyond the letter of the law to new, creative pollution prevention approaches.

3. *What Components of the Environment Are Affected, and How?*

The brewing process

The environment we are concerned with in this case is primarily the industrial enviroment within a brewery and its supporting facilities. Beer has been made in essentially the same way for more than 6,000 years, in the Babylonian, Chinese, Egyptian, Greek, and Roman civilizations, as well as in most modern societies. The basic process is quite simple: malted barley, or a mixture of grains is heated ("mashed") and then mixed with water to convert starches to sugars. The grain is then removed and the remaining liquid boiled with hops (a plant product). Yeast is added to encourage fermentation, in which sugars are converted into alcohol. Then the beer is aged, filtered and bottled. Different types of beer use different combinations of hops and malt, and refinements in the brewing process yield different carbohydrate levels (regular vs. light beers) and different alcohol contents.

A typical modern brewery consists of a number of unit processes to achieve finished beer, as shown in the diagram below. As in most industries, breweries have a special language for their processes. A brief glossary appears on the following page.

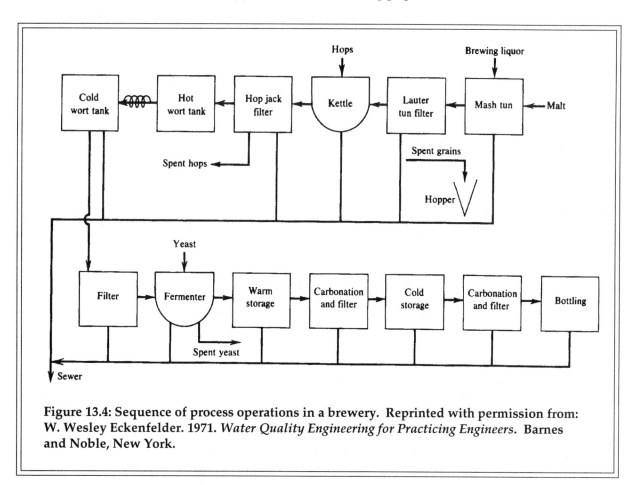

Figure 13.4: Sequence of process operations in a brewery. Reprinted with permission from: W. Wesley Eckenfelder. 1971. *Water Quality Engineering for Practicing Engineers.* **Barnes and Noble, New York.**

Unit processes and terminology in brewing

Mash: the mixture of water and malted barley and other grains such as corn and rice; the starting point for beer-making.

Mash tun: the vessel in which mash is heated with water to encourage enzymatic changes in the mixture.

Lauter tun: the vessel in which the mash is allowed to settle and in which screening occurs. "Spent" (used or waste) grains from the lauter tun are traditionally removed to a storage container for later disposal.

Brewing kettle (or "copper"): the vessel in which the mash is mixed with hops and boiled.

Wort: the liquor in the brewing kettle; the mixture of boiled mash and hops.

Hopjack filter: the filter used to remove spent (used) hops from the wort.

Hot wort tank: a holding tank where the hot wort is cooled.

Cold wort tank: a storage tank for the cooled wort.

Filter: the cold wort is passed through a filter stage, sometimes made up of diatomaceous earth, to remove residual particles and clarify the liquid.

Fermentation tank: the tank in which yeast is added to the filtered wort and in which fermentation takes place, usually at about 16 to 20 degrees Celsius.

Warm storage: the tank in which fermented beer is stored before further filtration.

Carbonation and filtration: in this step the beer is filtered further and then carbonated to an acceptable level of effervescence.

Cold storage: the filtered, carbonated beer is stored at cold temperatures for a time to improve its taste.

(Re)carbonation and filtration: following cold storage, the beer is filtered and carbonated again.

Packaging: finished beer is bottled in glass bottles, canned in aluminum cans, or packaged in kegs for final sale.

Waste sources and volumes

The various unit processes involved in brewing create a variety of wastes. The diagram shows emissions of spent grains, hops, and yeast. Less obvious are releases of washwater from many processes to the sewer system (which in most cases ultimately goes to a municipal wastewater treatment plant or similar facility). Air emissions, for instance of VOCs, can also be a problem, as Coors discovered to their cost. And solid waste—packaging from raw grains and

Table 13.1 Wastewater volumes in a brewery

Brewhouse (mash, lauter, kettle, hop jack, worts)	25%
Yeast and fermentor	3%
Stock house finishing	7%
Cold storage, cooling	4%
Bottling, filling, pasteurizing	60%

Note: a typical barrel of beer generates about 370 U.S. gallons or about 1,400 L of wastewater.

hops, used cans and bottles, and so on—creates another "invisible" waste stream.

In traditional brewing, spent grains can account for about 5.5 kg (dry weight) of waste per barrel of beer brewed, where a barrel is 31.5 U.S. gallons or 119.2 liters. Spent hops make up another 0.22 kg (dry weight) per barrel. Surplus yeast and waste products from the fermentation process amount to about 0.24 kg (dry weight) per barrel, and filter cake (diatomaceous earth, proteins, yeast cells, and hop resins) makes up another 0.454 kg (dry weight) per barrel.

In terms of water pollution, a major contaminant of concern in brewing is BOD (biochemical oxygen demand)—a parameter that measures the oxgyen demand of decaying organic material in the effluent. Suspended solids (particulates) are also problematic, because of the several mixing/mashing steps that allow solids to break down into smaller particles. Most brewery wastes have a BOD:N:P (BOD to nitrogen to phosphorus) ratio of 100:4:1, which makes them relatively nutrient-poor and thus difficult to use in biological treatment processes.

Coors' industrial ecology

Coors has identified a number of key waste areas in its brewery operations. These include spent grains, can waste, wood waste (from shipping pallets), and solvent use. They have actively sought means of reducing these wastes while developing alternative products and secondary uses. Some of their initiatives include:

Golden Aluminum

This subsidiary of Coors Brewing buys between 90 and 100% of the cans sold by Coors. At 70% recycled materials, Coors cans now have the highest recycled content in the industry. (Coors also employs continuous casting processes for its cans and uses superlight casting to minimize raw material use in its can production.) In 1955, Coors was a leader in introducing the aluminum can as an advance on the steel can. Its 50% lighter weight means much lower transportation and other weight-sensitive costs.

ZeaGen

ZeaGen is another Coors subsidiary that Coors started to make best use of its spent grain products. ZeaGen uses spent grains and other materials to manufacture animal feed, fertilizer, food supplements, health foods, and a variety of other products.

Bio-T and Golden Technologies

Reacting to concerns about the environmental and human health impacts of petroleum-based solvents, Coors developed a citrus-based solvent called Bio-T, now used widely in its operations. The solvent is now sold outside the company through another spin-off subsidiary called Golden Technologies.

Coors Ceramics and ACX Technologies

In an effort to reduce energy costs in brewing, Coors developed advanced ceramic materials for its brewing vessels. Coors Ceramics is now a division of ACX Technologies and develops super-lightweight engine parts and ceramic filters for cleaning air and water.

Composting

Wooden shipping pallets used to make up half the brewery's waste. (Some estimates suggest that a typical shipping pallet is used about 1.5 times before it must be discarded.) Coors now employs more durable pallets and composts spent grains with shredded pallets to make a compost product. Overall wastes from the brewery were reduced 30% in this step.

Some information on Coors' progress with these activities is illustrated in Figures 13.5-13.7 and Table 13.2.

Coors now has a broad plan to reduce waste across the company, including reduction of packaging waste, reuse, source reduction, recycling, improved recyclability and composting. Among its goals are to:

- increase glass recycled content to over 35%
- increase aluminum recycled content to 75%
- increase corrugated cardboard recycled content to 90%
- have 20% recycle content in clay-coated secondary packaging
- have 15% recycled content in composite can packages
- continue to compost scrap wood
- reduce all packaging at source where functionally feasible
- expand internal programs for recycling paper, plastic, glass, and aluminum
- maintain 90%+ recycled content in point-of-sales materials

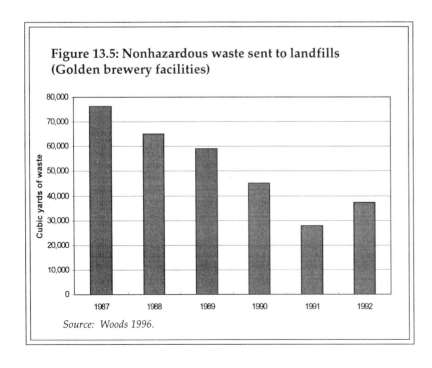

Figure 13.5: Nonhazardous waste sent to landfills (Golden brewery facilities)

Source: Woods 1996.

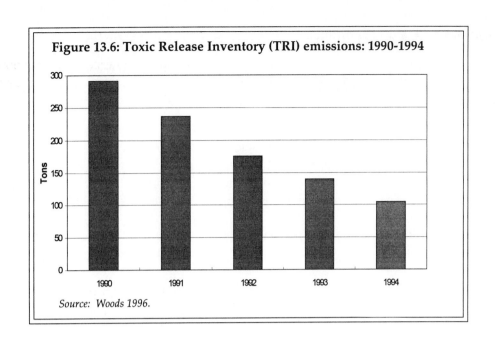

Figure 13.6: Toxic Release Inventory (TRI) emissions: 1990-1994

Source: Woods 1996.

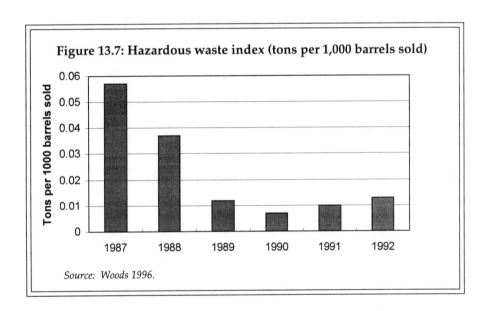

Figure 13.7: Hazardous waste index (tons per 1,000 barrels sold)

Source: Woods 1996.

4. How Can I Analyze This Information?

In part because of its corporate philosophy, Coors has always welcomed innovative approaches to waste minimization. Until the late 1980s, these approaches tended to evolve on an *ad hoc* basis, without a formal plan or structure. In 1988, Coors decided that a more systematic approach, implemented company-wide, would be better for improving the environmental and economic performance of the company. They envisioned four elements to such an approach:

1. *Corporatewide environmental principles establishing zero-resource budgeting*—ways of driving pollution and waste towards zero while achieving a net cost savings

2. *Environmental audits* to identify sources and characteristics of waste and options to reduce or eliminate it

3. *Decentralized management systems* that make every employee responsible for waste, and give individual employees the authority to reduce waste under their control

4. *Full environmental accounting* using two methods to track the purchase, use, and waste of materials and assign the costs of waste to individual profit centers

In a sense, Coors established a system that used waste as a corporate indicator of inefficiency and insufficient technology—a companywide expectation that each and every employee will do his or her utmost to identify sources of wastes and find ways to reduce them.

Probably the most powerful tool available for a waste reduction program such as Coors' is the waste audit—a systematic and thorough examination of every component of the process with respect to key indicators. Coors has selected three indicators for a first assessment of their progress: a solid (nonhazardous) waste index (tons of solid waste per 1,000 barrels of beer sold), a toxic release index (tons of toxic release inventory emissions per 1,000 barrels sold), and a hazardous waste index (tons per 1,000 barrels sold). Each of these indicators has shown encouraging reductions since 1987 (see Figures 13.5-13.7), giving the company further incentive to continue and expand its programs and develop its waste minimization subsidiary operations.

Other powerful tools include full environmental cost accounting, so that waste disposal and pollution costs can be assessed on a profit-center basis. Employees are held personally responsible for these costs, as they would be for other costs of doing business, so there is good incentive from a profit perspective to reduce wastes and find profitable alternatives.

There is no question that Coors' deep corporate commitment to waste

Table 13.2: Coors recycling data: 1992	
Typical recycled content (post-consumer):	
Glass	30% (106,100 tons)
Corrugated cardboard	60% (25,389 tons)
Aluminum	68% (69,615 tons)
Reuse:	
Glass	84,318 tons
Stainless steel kegs	71,799 tons
Keg boards	2,964 tons
Pull sheets (reprocessed pallet replacements)	7,351 tons
Recovery and recycle:	
Corrugated cardboard	4,797 tons
Office waste paper	126 tons
Plastic film	300 tons
Composting (scrap wood)	1,474 tons

Source: Woods 1996.

minimization has influenced all parts of the corporation and every employee. Without this central commitment, success would have been harder to achieve. Hand in hand with corporate commitment comes individual commitment, however. Coors' decentralized management structure may seem risky and unfamiliar to some more traditional businesses, but it has been very important in allowing employees the freedom to take chances and exercise creativity in waste reduction.

Box 13.3: Tips on the waste audit

The purpose of a waste audit is, very simply, to account for all the sources and sinks of waste in a given operation. In practice, this can be daunting, especially for the novice. Here are some practical tips:

1. *Begin with a thorough understanding of how the process works. What unit processes are involved? What raw materials or feedstocks are employed? What products are generated? What by-products?*

2. *Determine the physical layout of the plant. Usually, plant personnel will be able to supply a floor plan so you can see the proximity of one unit process to another. Where do raw materials enter the facility? Where do products leave? Can you map the locations of discharges to air and water?*

3. *Choose an indicator or indicators of interest to you: don't try to examine all variables at once! To do so would simply confuse you and waste time, because not all unit processes will be equally important for all indicators.*

4. *Develop a schematic drawing of the sequence of unit processes. Consider each unit process separately, one at a time. If a process has little or no relevance to your indicators, use a green marker to color that box. If the process is of moderate importance, color it yellow. If it's a real "hot spot" for your indicator, color it red. Confirm your assessment with plant personnel.*

5. *Now go back and reexamine your unit processes in reverse order, looking at the most important ones first. Can you estimate for each what mass of your material enters the process? Leaves the process? Is lost or generated within the process? Plant personnel may be able to help you with some of this information, but you may also have to collect new data in the plant.*

6. *Attempt a mass balance: if you know that a certain mass of material is entering the plant (for example, mass of aluminum), can you account for that mass either inside the plant or as waste leaving the plant? Don't expect great accuracy here, but even the attempt will likely show you where you need to do more work. It's quite common to "lose" mass in the plant— at least on paper! In other words, you may not be able to account for all of the mass entering the plant. Is it possible that some of the missing mass is lost as leaks or spills? Fugitive emissions (e.g., leaky chimneys)? You may have to refine this analysis over a period of time as you collect more data.*

7. *Where is most of the waste generated? What unit processes are most important in contributing this waste? What waste streams (solid, liquid, air) are most important? What remedial measures might be useful in reducing waste (use the pollution prevention heirarchy as a guide). What impact will these measures likely have on total waste generated and on waste disposal costs?*

5. *How Can I Use My Findings to Reach a Solution?*

Coors' experience is an excellent example of how corporate commitment and employee creativity can combine to reduce human impact on the environment. Coors' actions have included not only technological solutions but also management restructuring and full environmental accounting. It may be difficult to replicate their success without these features in place.

1.   What is the problem?

In Section 2, we identified the problem as how to reduce pollution effectively and efficiently without creating other pollution problems or resulting in a net cost to the company.

2.   In what ways do human activities have impact on the natural environment to cause "a problem"? How do these mechanisms give you clues to possible solutions?

Most manufacturing operations use raw materials and discard solid, liquid, and gaseous wastes. It is only in the last twenty years or so that manufacturers have become aware that these wastes actually represent lost profits. In fact, poor "housekeeping" and less-than-optimal operation of process equipment probably contribute most of the wastes leaving a manufacturing operation. So human impacts on the environment may in this case relate as much to careless operation as to unavoidable by-products of manufacturing. This suggests (as it has to companies like 3M and Coors) that preventing pollution before it occurs can save money as well as reduce environmental impact.

3.   What governments are responsible for the issue? Whose laws may apply?

In the U.S., the Environmental Protection Agency has overriding authority for pollution control activities, although some of the implementation of EPA's rules is delegated to the state governments, often through enabling state law; both federal and state governments would therefore be involved in this case.

4.   Who has a stake in the problem? Who should be involved in making decisions?

Coors is clearly the most prominent stakeholder here, although the governments listed in step 3 would clearly have a regulatory and enforcement role. Coors' customers may be interested in the company's environmental practices and would certainly bear the brunt of any cost increases that resulted from Coors' environmental management practices. Other possibilities would include transportation companies carrying Coors products and retail outlets where the products are sold.

5.   In the view of your decision-making group, what are the attributes of a satisfactory solution? In other words, when will you be satisfied that the problem is "solved"?

Coors uses three indicators of environmental progress: a nonhazardous solid waste index, a toxic release index, and a hazardous waste index, each of which is measured in tons per 1,000 barrels of beer sold. Cost is probably an implicit consideration as well. Other measures of performance could include targets for individual pollutants or materials, although given the complexity of Coors' operation it probably makes sense to combine individual measures into composite indices, as Coors has done.

6. *How will you evaluate (test, compare) potential solutions?*

You could evaluate the performance of this industrial ecosystem in the same way that the Hubbard Brook ecosystem could be assessed (see Case Study 11), using computer simulation, multiattribute decision analysis, cost-benefit analysis, or other tools. Coors has chosen to use a fairly simple comparison of its three key indices, looking for a constant downward trend in waste production and steady progress toward their recycling and other environmental targets.

7. *What are all the feasible solutions to the problem?*

The pollution prevention literature is remarkable for its creative solutions to problems that were once thought to be unavoidable. Feasible solutions for waste management at a facility like Coors' Golden brewery could certainly be drawn from this literature, but equally well could come—as they have at Coors—from the ideas and insights of ordinary people.

8. *Which solutions work "best" in terms of the attributes you identified in (5)?*

From Coors' perspective, good solutions are those that make progress towards the corporate recycling targets and provide a steady reduction in waste emissions. In this case, it may make sense to use a combination of all feasible measures (if financial resources allow) to obtain the maximum possible waste reduction.

9. *Which solution will be easiest to implement?*

Experience has shown that high-cost technologies can be easier to implement than measures that require people to change their behavior. For example, it may be easier to install a costly water recycling system than to ensure that a worker limits the amount of water used for washing floors and equipment. Education programs can go a long way toward increasing worker acceptance of change. Coors' corporatewide pollution prevention strategy places the onus on every worker to track and report wastes and to find ways to reduce. Employees are held personally responsible for the costs of waste disposal and pollution—a powerful incentive for employees to avoid waste.

10. *What steps are needed for successful implementation? Who will pay? Who will monitor progress?*

Implementation requirements will depend on the individual measures identified. Coors' experience has shown that pilot projects are a useful way to test a new waste utilization technology: if the pilot works, the test can be expanded to full scale. Waste management costs at Coors are assessed on a profit center basis, so implementation plans should certainly identify individual profit center (and perhaps worker) responsibilities, projected costs, and timing of change. Ongoing monitoring and reporting is an important part of most pollution-prevention plans, because it provides evidence of progress or reveals the need for further action.

6. *Where Can I Learn More About Pollution Prevention and the Brewing Industry?*

The following sources contain a variety of information on pollution prevention, brewing, and the Coors experience.

W. Wesley Eckenfelder. 1971. *Water Quality Engineering for Practicing Engineers.* Barnes and Noble, New York.

P. Fellows. 1988. *Food Processing Technology: Principles and Practice.* Ellis Horwood, Chichester, U.K.

U.S. Environmental Protection Agency. 1993. Design for the Environment: Environmental Accounting and Capital Budgeting Project Update #1. EPA Publication 742-F-93-007. Prepared by the Office of Pollution Prevention and Toxics, Washington, D.C.

U.S. Environmental Protection Agency. 1994. Fact Sheet on Pollution Prevention Financial Analysis Software. EPA Publication 742-F-94-003. Prepared by the Office of Pollution Prevention and Toxics, Washington, D.C.

U.S. Congress Office of Technology Assessment. 1992. Green Products by Design: Choices for a Cleaner Environment. Publication E-541, Washington, D.C.

Owen P. Ward. 1989. *Fermentation Biotechnology: Principles, Processes and Products.* Prentice Hall, Englewood Cliffs, New Jersey.

S. Woods. 1996. Coors' ten ways to prevent pollution by design. In: J. Fiksel (ed.), *Design for Environment.* McGraw-Hill, New York.

The following chemical engineering encyclopedias may also be useful in understanding the brewing process:

Kirk-Othmer Concise Encyclopedia of Chemical Technology. Wiley, New York, 1985.

Perry's Chemical Engineer's Handbook, 6th Edition. McGraw-Hill, New York, 1984.

"How can we reconcile human needs with the need to preserve biodiversity in a national park?"

1. What Is the Background?

A rich biodiversity heritage

Although it ranks only twenty-third among African countries in terms of its areal extent, Cameroon has the fifth highest number of mammal and plant species on the continent, including more than 40 globally threatened species. Its rich biodiversity has made it the focus of numerous conservation projects, among which is Korup National Park, covering an area of about 126,000 ha on the country's western border. Much of the park's humid lowland forest has never been logged, and these old forests now support more than 250 species of birds, more than 400 species of trees, and one-quarter of all the primate species found in Africa. The area is particularly noted for its high number of endemic species—species found only in that region.

There are many pressures on this park, among them deforestation. Most of western Africa's great forests have now been cut and converted to plantation agriculture. In central Africa, the original forests are still mostly intact. Cameroon's position between the two regions makes it a zone of tension for resource extraction activities like forestry. The country has a system of protected areas, but among them are many seriously degraded "reserves" and only one park—Korup—with a formal management plan.

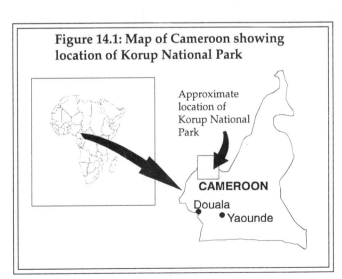

Figure 14.1: Map of Cameroon showing location of Korup National Park

Approximate location of Korup National Park

CAMEROON

Douala

• Yaounde

In 1989, the Cameroonian government reached an agreement with the World Wide Fund for Nature (WWF), who now, in cooperation with the United States arm of WWF and the Overseas Development Agency, manages the park. A variety of biodiversity research projects and training programs are ongoing within the park, and additional research-linked funds (for instance from the U.S. Department of Defense) have been granted in support of biodiversity preservation there.

As part of the long-term management of the park, and as required by the legal agreement between Cameroon and WWF, the thousand or more people currently living the park must be resettled outside its boundaries. This resettlement has not yet occurred, but the people still living in the park are no longer permitted to hunt. Instead, they scratch out a meager farming income, take charity, and engage in various criminal pursuits like smuggling and poaching to recover lost income.

The six communities inside the park are now greatly impoverished, lacking both food and the means to buy food. Soon after the management agreement was signed, the park ran out of money for promised roads and services for the villagers, so it is not clear when resettlement can begin or how it can be accomplished fairly and equitably. In the meantime, WWF says that the continued presence of the villages inside the park threatens the preservation of biodiversity and breaks the legal agreement on park management. Yet WWF managers are the first to admit that they seriously underestimated the costs of resettlement. Even though $10 million has been spent to date on roads (the most expensive item) and other services, there is much more to do, including construction of at least two promised roads.

> **Box 14.1: The benefits of biodiversity**
>
> *Although environmentalists are quick to voice the need for biodiversity preservation, the practical implications of preservation are not always clear even to scientists. In Korup National Park, a recent discovery demonstrates these benefits in a very practical way.*
>
> *In the early 1990s, an obscure vine growing in Korup National Park was identified as containing the chemical michellamine B, which appears to inhibit the HIV virus that causes AIDS.*
>
> *The vine, a member of the genus* Ancistrocladus, *has not been found anywhere else in the world. It is one of only three plants being intensively tested by the U.S. National Cancer Institute as a potential AIDS cure.*

2. *What Problem Are We Trying to Solve?*

In some primitive societies, hunters use simple tools and weapons that limit their hunting efficiency and therefore their impact on populations of their prey. In Korup National Park, however, villagers hunt with modern weapons (in the case of elephants, they use submachine guns), and their hunting serves not just family food needs, but also domestic and export market demand. WWF managers say that the impacts of this hunting are not benign but have significantly reduced the numbers of certain animal species over the past 10 years. Villagers believe that hunting has always been their right and that there will always be enough animals to hunt in the park.

Despite the villagers' belief in the productivity of the forest, there is general agreement that the dispute must be settled and strong motivation for early settlement in the form of the legally binding park management agreement. In the long term, it seems clear that the park's status as a biodiversity reserve can be assured only if resettlement is accomplished, and accomplished peacably and without economic detriment to the villagers.

Our problem, in this case, is therefore to determine the best way to resettle the villagers, including community education, provision of adequate dwellings and services (especially roads), and related considerations.

Box 14.2: Game animals valued for food

Savannah lands	Savannah and forests	Forests
Cheetah	Baboon	Chimpanzee
Rhinoceros	Lion	Mangabey
Giraffe	Leopard	Gorilla
Warthog	Hippopotamus	Colobus monkey
Giant eland	Sitatunga	Drill
Roan antelope	Bushbuck	Mandrill
Buffon kob	Buffalo	Bush pig
Hartebeest	Elephant	Bongo
Topi	Hyena	Porcupine
Gazelle	Serval cat	
Patus monkey	Civet cat	
	Reedbuck	
	Duiker	
	Aardvark	
	Waterbuck	
	Oribi	
	Tantalus	
	Crocodile	

Source: Balinga 1978.

---

3. What Components of the Environment Are Affected, and How?

Major ethnic groups

There are two major ethnic groups in the Korup region. The eastern part of the park is inhabited by Bantu tribes, including the Bima, the Bakoko, and the Ngolo. These peoples grow coffee, cocoa, avocado, and mango on rich volcanic soils in that region. Their cultivation techniques are largely traditional and have had little impact on the park ecosystem.

To the west, the population is non-Bantu. One tribe, the Korup (who, like the park, are named after the river that runs through the area) make up a little less than two-thirds of the total park population, or about 650 people. The soils in their area are much less fertile than those farmed by the Bantu, so the Korup and the Ejagham, the other non-Bantu tribe in the area, have traditionally lived as hunters and fishers. A lively trade in "bush meat"—meat from wild animals such as chimpanzees—has grown up in the area, centered on the main Korup village of Erat. Korup National Park is located right on the Nigerian border, and many Korup in fact live in Nigeria. Smuggling of various plant and animal products across the Nigeria-Cameroon border has become a major source of income for people in this area.

For the villagers, one of the biggest challenges of forest life is transportation. some settlements are located 30 or 40 km from larger urban centers, so goods destined for markets, the sick and injured, and all nonnative materials must currently be brought on foot over many miles of footpath. For much of the year, these footpaths are difficult to traverse—the region receives between 5 and 10 m of rain a year, and walkways are often

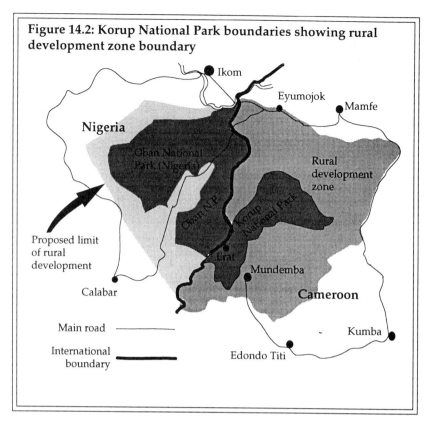

Figure 14.2: Korup National Park boundaries showing rural development zone boundary

Ikom

Eyumojok

Mamfe

Nigeria

Oban National Park (Nigeria)

Rural development zone

Oban N.P.

Korup National Park

Proposed limit of rural development

Erat

Mundemba

Calabar

Cameroon

Main road ————

International boundary ▬▬▬▬

Edondo Titi

Kumba

awash in mud. This has been a particular problem for access to the health clinic located some 15 km from the villages. Poor access is therefore one of the main reasons why good roads are so attractive to the villagers, and why good roads are one of the few incentives park managers can offer these people to leave their traditional homes.

The buffer system

Korup National Park is the only one of Cameroon's natural reserves to have a formal management plan. The plan calls for a system of "buffer zones"—zones of good agricultural productivity adjacent to park areas. The idea behind buffer zones is that they will protect sensitive species in the park from human interference, and also offer a location for resettlement of park residents in an area not too distant from their original villages. Much of the agricultural land to the south and east of the park is highly fertile, and although park managers may not be able to persuade all poachers to convert to farming, the productivity of the land may encourage some at least to sign agreements not to hunt.

Land beyond the buffer zone, and indeed some within the zone, is forest under concession to loggers. Environmentalists believe that rapid careless cutting of this forest could have impacts on the hydrologic regime in the Korup River and its tributaries, thus affecting both agricultural activities in the buffer zone and conservation efforts in the park.

Box 14.3: The African hunting tradition

In Cameroon, as in most of west and central Africa, wildlife is by tradition regarded as a regular and God-given source of protein. Our people believe that hunting should be free and cannot understand why they should be restricted in this activity. After all, they have always hunted since the time of our forefathers. It should not be surprising, therefore, that despite legislative measures, poaching has been the order of the day for years and is entrenched among the habits of the people.

Victor S. Balinga
former member of the staff
Direction des eaux et forets et chasses
Republic of Cameroon
1978

The Cameroonian government has sold many forest concessions in recent years and receives more than $100 million a year in forest taxes. Concessions are currently held by corporations from several nations including France and Germany.

Future development

Future development in Korup National Park will rest with park managers, and in particular with WWF and its partners. If adequate services cannot be provided for villagers outside the park, there may be other options for resolving the problem. The Cameroonian government has taken the position that mandatory resettlement is to be avoided. An alternative would be to restrict hunting to sustainable levels, to limit species taken, or weapons used. Clearly, there would be major problems in enforcing such a regime, particularly with the access problems already existing in the park.

In fact, many villagers want to move. They are attracted by the prospect of better health care and education. One of the effects of the presence of Westerners in the area has been to raise villagers' awareness of the value of education. Andrew Allo, the Cameroonian manager of the park project, believes that when villagers understand the need to move, when they are given the services they want, and (perhaps most important) when they are involved in developing their own new villages, they will accept and even welcome the idea of resettlement. Allo, the son of a local tribal chief, believes that many African resettlement schemes have failed because of arbitrary or peremptory moves that do not allow sufficient time and assistance for villagers to make a new life.

> *"I ask people what they want for their villages and they all say, a road, a school, a clinic and a future."*
>
> *Andrew Allo*
> *Cameroonian manager of the*
> *Korup National Park Project*

Among the features that must be included in successful development of new villages are the infrastructure for agriculture. Facilities such as seedling nurseries will be an essential adjunct to village, road and field development, especially for poachers-turned-farmers.

Yet this "carrot" approach to resettlement should not be expected to convince people of the benefits of conservation. Some critics of resettlement programs say that government-led

---

Box 14.4: The Convention on Biological Diversity

The U.N. Conference on Environment and Development, held in Rio de Janeiro in June 1992, provided an important forum for the discussion of global environmental issues. The United Nations Environment Program assisted in the drafting of a Convention on Biological Diversity, which was signed by 156 nations during the conference.

The Convention has the goal of protecting biodiversity and restoring damaged ecosystems. It also contains provisions for the equitable distribution of benefits from the research and development of genetic resources, so that profits are shared between the developer and the nation from whom the resource originally came.

At present, the Convention is gaining momentum as more and more governments agree to sign it. As part of this momentum, a number of specialty conferences have been held in the years following the Rio conference. In addition, a Global Environment Facility is under develement by the U.N. and the World Bank to provide loans for projects that have environmental benefits in preserving biodiversity and maintaining natural habitats, which reduce the emission of greenhouse gases, which stop pollution of international waters, and which protect the ozone layer.

programs of this type allow regulatory and technical agencies to control populations and resources closely. They may work against the development of village autonomy and dignity, and indirectly against sustainable use of natural resources.

Funding sources

Resettlement has been one of the largest costs of the Korup National Park Project. More than $10 million U.S. has already been spent on the project, with roads the most costly item. WWF has agreed to fund the development of new villages for resettlement, but WWF acknowledges that actual costs have far outstripped original estimates. In fact, some managers believe that a proposal containing the real costs of the project would have been rejected by Fund officials.

In recent years, WWF has sought external funding to complete the resettlement project. After 5 years of searching, the European Economic Community committed to a donation of about $3.77 million U.S. to fund the construction of a single road servicing three new villages outside the park. The U.S. Department of Defense has contributed a further $800,000 for bridge construction. A second road is still needed, however, and the central Korup village of Erat—headquarters of the poaching trade—cannot be resettled until money is available to build that road. WWF continues to seek donors for the project through an extensive public information program.

Jurisdictional issues

The Korup National Park Project involves cooperation among a number of groups, both government departments and nongovernment organizations. National agencies with a particular interest in the project include the Secretariat of State for Tourism, the Ministry of Planning, and the Directorates of Forestry and the Environment of the Ministry of Agriculture. The National Office in charge of forest development is also interested in the project.

Nongovernment conservation organizations involved in the project include WWF-U.K., WWF-U.S., Wildlife Conservation International (U.S.), the International Council on Bird Preservation, Living Earth (U.K.), and the Fauna and Flora Preservation Society (U.K.). In Cameroon, local groups include WWF-Cameroon, Enviro-Protect, and the Club des Amis de la Nature.

The vast complexity of this stakeholder network—even excluding the villagers themselves!—creates a further layer of confusion, delay, and repetition even in the most well-intentioned project activities.

4. How Can I Analyze This Information?

This case does not require formal "analysis" as much as it requires sensitivity to the local culture and traditions, understanding of the complex administrative structures for the park and the country, and the needs and wants of diverse stakeholders.

We can identify several methods that are nevertheless important in managing this issue.

Education

Steve Gartlan, WWF manager for the Korup National Park Project and a Korup researcher for almost 20 years, believes that education of the villagers is central to successful park management. At present, awareness levels within the villages are low. The Korup people are drawn by the prospect of wealth, which they believe will follow resettlement with all the amenities—electricity, piped water, health clinics, and schools, and roads to markets—that it will bring. The villagers distrust the project managers, possibly because of broken promises about roads and infrastructure, and believe that they have so far gained nothing from the park project.

Villagers also lack an understanding of why Westerners value the rainforest so highly that they would displace people to preserve it. Although WWF managers and others have held many meetings with the villagers, older adults still claim not to understand their message. Education of young children through the village schools has been more successful, but criticism from chiefs has made some school teachers secretive about what they are teaching.

A successful education program is probably essential in the long-term reduction of poaching and smuggling. Some project managers are sponsoring village children in attending secondary schools distant from Korup. Although this will clearly have benefits for park preservation, it may have detrimental effects for the preservation of traditional cultures. These forces make it particularly important that African values, as well as Western values, form part of the education prcoess (see next section).

Managing for whom?

In a recent article, Joseph Zano Zvapera Matowanyika notes that government policies are usually portrayed as being "value-free". In fact, these policies may respond to a narrow segment of the population while ignoring the needs of peasants who inhabit the land to be managed. A century of paternalistic colonial bureaucracy has contributed to a policy-making culture that is largely alien to the African population. Matowanyika believes that even the "desperate needs" of local residents are often treated as irrelevant or somehow outside the realm of bureaucratic management. As a result, a majority of Africans may view wildlife management practices, and the people that employ them, as foreign and uncaring.

> "The process of wildlife management is portrayed as a value-free matter and the regulations set up are deemed rational and objective. The concerns, interests and even the desperate needs of local residents are often treated as irrelevancies, outside the system, outside the scope of management considerations."
>
> *Joseph Zano Zvapera Matowanyika*

At issue here is a conflict of values and priorities in how best to use a shared resource. Unlike stakeholders in more developed countries (see for example Case Study 4, on New Zealand, and Case Study 7, on the United Kingdom), African villagers have neither the knowledge nor the social status to participate in collaborative decision making.

Other issues

This case demonstrates the complexity of political and nongovernment interests in a major resource protection scheme, the lack of funding available even for basic infrastructure, and the conflicts of values and priorities that occur when different cultures interact. Not as obvious from this discussion are the shortage of trained staff (see also Case Study 9, on Costa Rica), the vast international market for the products of poaching and smuggling, and the poor knowledge base about the dynamics of savannah and tropical forest ecosystems.

This case has emphasized the human components of the Korup National Park management issue. It emphasizes the need for consideration of local values and local conditions, and thus implicitly suggests the need for local planning and conflict resolution.

As a remedy for this, some studies of the Korup National Park and other similar areas in Africa and elsewhere have begun to collect data systematically on not only on biophysical resources but also on cultural issues, community history, perceived conflicts, present and past land uses, and traditional technologies. This information can be invaluable in tailoring management schemes to local conditions.

It goes without saying that planning activities such as those for park management and village resettlement should involve not only local values but local people, local languages, and local cultural traditions. In this way, multiobjective management frameworks can be developed with the cooperation and understanding of those they will affect most directly. In the longer term, such joint planning initiatives can form the basis for future joint management of park resources.

5. How Can I Use My Findings to Reach a Solution?

The management of Korup National Park, and the situation of the villagers living in it, is one of the more difficult cases presented in this book. Tradition and quality of life are threatened by international commitments, some of them legally binding. There will be no easy solution for Korup, but using the problem-solving framework from the Introduction may help you sort out some reasonable places to start.

1.   What is the problem?

In Section 2, we identified the problem as determining the best way to resettle the villagers of Korup National Park, including community education, provision of adequate dwellings and services, and related considerations.

2.   In what ways do human activities have impact on the natural environment to cause "a problem"? How do these mechanisms give you clues to possible solutions?

In Korup National Park, a richly diverse and as-yet largely undisturbed ecosystem is threatened by excessive hunting pressure. Some of this pressure is driven by poaching and smuggling pressure; some of it simply provides food for the pot. The extent of the actual and projected hunting pressure is not well understood. It is clear that the values of the villagers differ from those of park managers and WWF staff, and presumably from those of government representatives. The villagers believe that unrestricted hunting in the park is their birthright; the others believe it will have serious consequences for pristine areas of the park.

3.   *What governments are responsible for the issue? Whose laws may apply?*

The Cameroon federal government is responsible for the park, which is part of the national park system. The government has made a legal commitment to WWF to relocate the villagers so that biodiversity in the park can be preserved, and we must assume this commitment to be binding. Other governments including the U.S. may provide funding to Korup through international development programs.

4.   *Who has a stake in the problem? Who should be involved in making decisions?*

Obvious stakeholders in Korup include the villagers, the park managers, representatives from WWF (in particular) and other environmental nongovernment organizations, and international development agencies. The buyers of smuggled goods and "bush meat" also have a direct interest in the outcome of this situation.

5.   *In the view of your decision-making group, what are the attributes of a satisfactory solution? In other words, when will you be satisfied that the problem is "solved"?*

Many villagers want to move, believing that they will find better health care, education, and roadways in the new areas. A satisfactory solution should therefore meet these needs at a minimum. It should also address the Cameroonian government's commitment to WWF, and thus indirectly WWF's interest in preserving biodiversity. It will be difficult to measure the extent of preservation except possibly as hectares of protected habitat. It will certainly be difficult to guarantee that lands set aside for preservation are in fact not hunted illegally. Finally, a satisfactory solution should be cost-effective and should have funding adequate sources clearly identified.

6.   *How will you evaluate (test, compare) potential solutions?*

A satisfactory resolution to this problem will almost certainly require consensus among the Cameroonian government, WWF, and any participating funding agencies. Although the participation of the villagers in decision-making is highly desirable, they have to date been largely excluded from the decision-making process. As a result, evaluation of potential solutions will likely, if not ideally, be conducted by these agencies using formal or informal consensus-building techniques (see Case Study 4, from New Zealand, for an example of these).

7.   *What are all the feasible solutions to the problem?*

There are probably innumerable solutions to this problem. The difficulty will be in finding adequate funding for them and securing agreement about them from the various agencies and the villagers. It may be helpful to review the literature on similar cases to determine what approaches have been successful in other circumstances. WWF staff believe that any successful solution will require an education component to raise the villagers' awareness of the impact of their actions. Such a program, if properly conducted, could also be useful in empowering the villagers to make their own planning decisions.

8.   Which solutions work "best" in terms of the attributes you identified in (5)?

This, like Case Study 2 on Bangkok, is a very complex issue to resolve and one that may take many years to reach a satisfactory conclusion. Clearly, a "best" solution will be affordable and will have the support of all the stakeholders listed in step 4. It should encompass ongoing surveillance and reporting to ensure that "protected" lands do not in fact deteriorate and to monitor the levels of poaching and smuggling in the park.

9.   Which solution will be easiest to implement?

The solution that will be easiest to implement will be the one that meets the villagers' needs (step 5), provides an alternative income source for poachers and smugglers (and an incentive to access that alternative source), and accommodates the concerns of WWF and the funding and regulatory agencies. Such a solution will be very difficult to find, primarily because of the immense funding obstacles.

10.   What steps are needed for successful implementation? Who will pay? Who will monitor progress?

The answer to this question is really the crux of this case study. In the preliminary stages of analysis, it may be enough to develop a list of potential funders, develop a consultation process, and prepare detailed cost estimates for the required health care, education and road facilities.

 # 5. *Where Can I Learn More About the Ecosystem, People, and Culture of Cameroon?*

The following sources are a useful introduction to the literature on multiobjective land-use planning in Africa and elsewhere.

P. Alpert. 1993. Conserving biodiversity in Cameroon. *Ambio* 22(1): 44-49.

V. S. Balinga. 1978. Competitive uses of wildlife. *Unasylva* 29(116): 22-25.

J. Z. Z. Matowanyika. 1989. Cast out of Eden: peasants versus wildlife policy in savanna Africa. *Alternatives* 16(1): 30-39.

H. R. Mishra. 1982. Balancing human needs and conservation in Nepal's Royal Chitwan Park. *Ambio* 11: 246-251.

M. Richardson. 1993. Wrestling with the preservation of the Korup rain forest. *Our Planet* 5(4): 4-7.

T. T. K. Tchamie. 1994. Learning from local hostility to protected areas in Togo. *Unasylva* 176(45): 22-27.

"How can we assess the true social and environmental impacts of a landfill?"

1. What Is the Background?

Landfilling: a common waste disposal option

Despite ongoing concerns about its social and enviromental impacts, landfilling remains the most common disposal option for municipal solid waste (MSW) in North America. One of the criticisms levied at landfilling is its high use of land. As land resources come under competing pressures from agriculture, urban development, recreational uses, and industry, the debate about how and where to site landfills has become more heated.

In recent years, the costs of designing, building, maintaining, and decommissioning landfills have increased dramatically, adding fuel to the fire. In the early 1990s, average tipping fees in the United States were about $26.50 a ton, with a range from zero to more than $200 a ton. The complexity of siting landfills is also increased by conflicting public agendas. A site that is hydrogeologically sound for landfilling may be located close to schools and residences. Landfills sited far from urban centers create increased transportation costs and higher environmental impact from the operation of heavy vehicles.

How can we make the best decisions about siting landfills, given this context of controversy and scientific uncertainty? One approach has been to evaluate all the "costs"—all the impacts—of a potential landfill project in terms

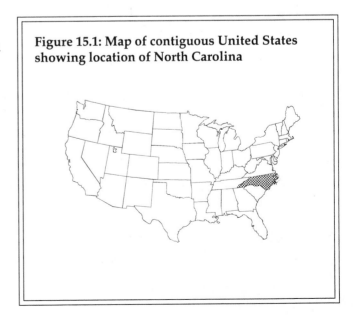

Figure 15.1: Map of contiguous United States showing location of North Carolina

of a single indicator: dollar value. While this approach has many critics, it does have value in simplicity and universality. Everyone, from public administrators to private homeowners, understands the meaning of a dollar and its relevance to their own interests. And since much of the debate around landfilling currently focuses on its costs relative to other options, assessing impacts in monetary terms may be a particularly valuable tool in the public debate. If we understand the true "costs" of a landfill, and the factors that were included in developing those costs, we should be able to make decisions that are more protective of the environment.

In this case, we examine a proposed MSW landfill for the city of Durham, North Carolina. Durham is a city of about 137,000 people—not large, but in combination with its neighboring city Raleigh (population 207,950), Raleigh-Durham is a major center for industrial and medical research in its Research Triangle Park. Durham, like many cities, has experienced a shortage of landfill space in recent years. A new landfill has been proposed for a 304 ha site, of which 81 ha would be used for actually landfilling. Initial tipping fee estimates, based on conventional cost estimation approaches, are for $32 U.S. per ton, or $35.20 per tonne. The tipping fee was calculated by Durham city staff and is intended to reflect the costs of land purchase, construction costs, operation and maintenance, and close-out. It does not explicity address the costs of environmental, social, or economic impacts that might be associated with the project.

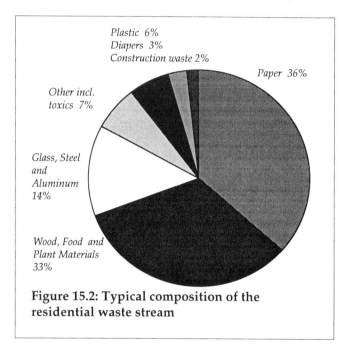

Figure 15.2: Typical composition of the residential waste stream

 ## 2. *What Problem Are We Trying to Solve?*

What do we mean by "impact"?

Landfills have many types of impacts, including biophysical impacts such as contamination of groundwater and socio-economic impacts such as those on property values. The first problem we encounter is therefore how to decide which potential impacts to include in an analysis and which to exclude.

Many analyses of the "costs" of landfilling consider only the obvious, direct costs, such as land purchase, construction costs, and supplies and labor for operation and maintenance. It can be argued that such an approach significantly underestimates the true costs of a landfill project. We know that landfills have the potential to create serious impacts on the built and natural environment, and on people who are part of those environments. Yet these impacts are almost always excluded from, for instance, a cost-benefit analysis of waste management alternatives.

This isn't surprising: it's very difficult to assign costs to intangibles like "quality of life". But it's not impossible. Techniques are available to allow environmental managers to itemize *all* impacts of a project, and to estimate the dollar value of those impacts.

How do we estimate "impact"?

The second problem we face therefore is how to make such estimates. In making a decision about estimation technique, we must realize that any choice we make will be controversial and any estimates we arrive at will contain error. Nevertheless, such an analysis can be a practical framework for organizing complex data, and a useful adjunct to more formal environmental impact assessment. Although reduction of diverse environmental impacts to dollars may require several analytical steps not easily understood by the lay community, the resulting assessment of dollar costs is easily understood by everyone in the debate and can therefore provide a useful basis for community discussions involving a range of stakeholders.

The problem we are trying to solve is therefore how best to assess the full range of social, economic, and environmental impacts of the proposed Durham landfill project, and to express that analysis in terms easily understood by the lay reader.

Box 15.1: Potential impacts of landfilling

Water pollution:	•*ground water pollution* •*surface water pollution*
Air pollution:	•*obnoxious odors* •*increased dust and particulates* •*methane production* •*carbon dioxide production* •*emission of other noxious gases such as hydrogen sulfide and volatile organics*
Noise pollution:	•*truck traffic* •*heavy earth-moving equipment*
Aesthetic insult:	•*poor appearance* •*litter* •*vermin and flies*
Land impacts:	•*soil erosion* •*alteration of drainage patterns*
Socio-economic impacts:	•*decreased quality of life* •*impacts on property values (present and future)* •*restriction on use of surrounding land*

Tipping fee: the fee charged to a landfill user for the disposal of a quantity of solid waste.

3. What Components of the Environment are affected, and how?

The Durham site

Before we can proceed to a more detailed analysis of impacts, it is important to start with a thorough understanding of the proposed site and its context. The actual configuration of the fill area on the site has not yet been determined. (A square configuration is used for the purposes of planning.) For now, the pending decision is whether this general area is even appropriate for a landfill, and if so, whether the true project costs, as reflected in an estimated tipping fee, are acceptable to the community. The population served is not explicitly considered in the planning for this project. Instead, an estimated filling rate is identified, based on current and projected waste generation rates. This rate could be higher or lower, depending on factors such as the level of community recycling or population growth.

Some basic assumptions for the site are given in Box 15.2. This case is described in more depth in Hirshfeld et al. (1992).

<table>
<tr><td colspan="2">Box 15.2: Basic assumptions, proposed Durham landfill project</td></tr>
<tr><td>Total land area available</td><td>304 ha</td></tr>
<tr><td>Fill area</td><td>81 ha</td></tr>
<tr><td>Average depth of fill (planned)</td><td>15 m</td></tr>
<tr><td>Depth of clay liner</td><td>0.6 m</td></tr>
<tr><td>Volume of cover</td><td>20% of total volume</td></tr>
<tr><td>In-place refuse density</td><td>590 kg/cubic meter</td></tr>
<tr><td>Land cost</td><td>$24,700/ha</td></tr>
<tr><td>Excavation cost</td><td>$6,175/ha*</td></tr>
<tr><td>Cost of routine landfill cover</td><td>$6,175/ha*</td></tr>
<tr><td>Average annual rainfall</td><td>102 cm</td></tr>
<tr><td>Leachate/precipitation ratio</td><td>0.4</td></tr>
<tr><td>Rate of filling</td><td>682 t/day</td></tr>
<tr><td>Period of public ownership
     following close-out</td><td>60 years</td></tr>
<tr><td>Rate of land appreciation</td><td>4% per year</td></tr>
<tr><td>Annual property tax rate</td><td>$1.50 per $100 assessed value</td></tr>
<tr><td>Typical value of residences
     within 5 km of site</td><td>$70,000</td></tr>
</table>

Source: Hirshfeld et al. 1992.
*Author's addition

Unit costs

In addition to information about the site size and filling rate, it is important to know the unit costs of various physical components of the project. The preliminary design for the site has a single composite liner, a leachate collection system, and a gas collection system. This would be a typical, but not a particularly "high tech" design approach for such a project. Leachate that is collected in the leachate collection system will be diverted to some off-site treatment facility. (Frequently, landfill leachate is sent to a municipal sewage treatment plant for further treatment.)

Some key unit costs are identified in Box 15.3.

Biophysical impacts

Depending on the individual site and proposed technology, a wide variety of environmental impacts may or may not occur. To the novice analyst, deciding which impacts to include can be a daunting task, especially where the system is an urban one and already considerably disrupted from its "natural" condition.

A few guidelines may help in identifying which elements of the environment are important to include in an analysis.

Examine the natural environment

Begin with an inventory of natural features: surface and groundwater resources (quality and quantity), air quality, noise levels, plant and animal species (especially rare or especially valued species), and human populations. Consider whether any subpopulations, for example pregnant women or children, are of particular concern for the project under consideration.

Examine the built environment

Human structures, machines, and activities also form part of the "environment" broadly defined. Make an inventory of these components of the system, paying special attention to heritage buildings, structures frequented by sensitive populations such as children, and similar features. Later in this section, there is a table showing the distribution of private homes in the area immediately surrounding the proposed landfill site; this is one example of the built environment.

Not all impacts need be negative

Sometimes, positive impacts on the environment occur as the result of human activities like landfilling. Don't ignore positive impacts and focus only on negative ones.

Identify key issues

As you begin to fill out your inventory of environmental features, certain elements or issues will begin to emerge as especially important. It's worth highlighting these early in the inventory because they will likely form the basis for more detailed analysis in later stages. They can also provide a good foundation for discussions with stakeholders.

Other considerations

Distribution of residences around the site

Although the precise orientation of the fill area is not yet known, it is possible to estimate the number of homes located within 0.5 km, within 0.5 to 1 km and within 1 to 5 km. Their distribution is approximately as follows:

Table 15.2: Distribution of residences around proposed landfill site

Relative location	Distance from site boundary (km)		
	Within 0.5	0.5 to 1	1 to 5
West	99	55	1
South	17	12	15
East	54	127	83
North	5	6	46
Total	*175*	*200*	*145*

Source: Hirshfeld et al. 1992.

179

Property values are known to depreciate more quickly closer to a landfill site. Work by Hirshfeld et al. (1992) suggests that $70,000 homes within 0.5 km of a landfill depreciate on average by about $18,000 after the landfill is sited; those between 0.5 and 1 km from the site depreciate by about $15,000; and those between 1 and 5 km depreciate by about $7,000.

4. How Can I Analyze This Information?

Calculating the true costs of a landfill project demands the use of several different tools. We can break the task up into the following steps.

Calculate the capital costs

The first, and perhaps the simplest, task is to calculate the costs of building the landfill. This step comprises several discrete calculations, including:

1. Calculate the total volume of the landfill (fill area x fill depth).

2. Calculate the portion of that volume that is usable for waste disposal (as compared to soil cover) = the volume of waste that the landfill can receive.

3. Calculate the mass of waste that the landfill can receive, from the known volume and estimated waste density.

4. Calculate the cost of buying the land for the site.

5. Calculate the cost of excavation.

6. Calculate the cost of a liner system, which will comprise four layers: a clay liner, a synthetic liner, a geotextile layer, and a drainage net. Use unit costs and known landfill volume.

7. Add the costs of a lift station and a leachate collection and treatment system.

8. Total these costs.

Calculate the operating costs

Operation and maintenance costs are typically much lower and less variable than one-time costs like construction. Assuming that the costs of hauling wastes are common to any waste disposal option, we can omit them from further consideration. Other operating costs can be calculated as follows:

1. Calculate the costs of covering waste with soil.

2.   Calculate the volume of leachate generated (rainfall/year x area x proportion of rainfall that yields leachate).

3.   Calculate the costs of treating leachate. This cost could be zero or it could reflect a lower sewer-use charge if diverted to a municipal sewage treatment plant. In this case, the city proposes an on-site treatment facility with known annual treatment costs.

4.   Calculate the costs of generating gases from the landfill. In this case, the city proposes to construct gas collection equipment as part of the landfill design, so there is no additional cost related to gas generation.

5.   Calculate the costs of any labor required to supervise these operations at the landfill (assume filling and treatment labor costs are included in the unit costs given). The level of supervision—and thus the costs of that supervision—are up to the analyst.

Calculate the costs of close-out

The costs of close-out, or decommissioning, are hard to estimate with accuracy, simply because we cannot guess what environmental or regulatory conditions may prevail 10 or 20 years from now when the landfill is full. An estimate of 5% of total capital (construction) costs is not unrealistic for site closure. Another 2% or so of construction costs should likely be allocated for ongoing site maintenance and monitoring after closure. These costs could go on indefinitely, depending on the "ownership" of the site in the post-closure period.

Calculate the costs of property-value impacts

Once the routine costs of construction and operation have been calculated, it should be possible to move into less traditional cost areas. Among these, and often foremost in the minds of site neighbors, are property-value impacts. We can never estimate these costs with certainty, but we can work from empirical evidence available in the literature, such as that produced by Hirshfeld et al. (1992). We would do this simply by calculating the number of neighboring homes, their distance from the site, and the probable depreciation in their property values, and sum those costs. In a sense, they are "one-time" costs because they occur once, following landfill siting, and are not likely to change with longer proximity to the site.

(A separate question is how we might use such costs once they have been estimated. One answer is that we could use them to estimate a higher tipping fee, which would result in higher revenues to the city from landfill operation. Extra revenues could then be used to compensate local homeowners for the decrease in their property values caused by the presence of the landfill.)

Estimate the costs of other environmental impacts

The "costs" of other environmental impacts are very difficult to estimate, and will likely be subject to considerable public debate. This doesn't mean that such estimates are not worth preparing or including in an anaylsis. Rather, it means that the analyst must make very clear what assumptions underlie the analysis and the methods used to estimate "costs" of these impacts. (Case Study 9, on ecotourism in Costa Rica, provides further discussion on this issue.) Analysts who do this usually act in accordance with the following steps.

Identify the probable impacts

People identify environmental impacts in different ways—this, in fact, is the crux of the debate on environmental-impact assessment. Some common approaches include subjective assessment (in which the analyst simply lists the impacts that he or she believes are important: a risky process because it inherently reflects the interests and biases of the analyst); and matrix analysis, in which the analyst uses one of several types of matrix to organize information about project activities and their impacts. Two example matrices are illustrated below. While matrix analysis is still highly subjective, it is often more comprehensive than a simple "eyeball" analysis. Its clear organization can also assist the analyst in seeking additional viewpoints from external stakeholders: the matrix can be circulated for discussion and revision, and its simple structure needs little explanation even for the lay person. By contrast, the rationale for a purely subjective analysis may be hard for the analyst to articulate and thus difficult to discuss with outsiders.

Example 1: Activity-component matrix

The most common form of an environmental-impact-assessment matrix is one like the following, where project activities (e.g., excavation, blasting, clay liner placement, waste hauling/truck traffic, and so on) are listed down the vertical axis of the matrix, and environmental components (e.g., surface and groundwater resources, air resources, wildlife habitat, and so on) are listed across the horizontal axis. The resulting matrix can then be filled with information about the direction and magnitude of expected impacts. Even though this analysis is necessar-

Figure 15.3: Example activity-component matrix

Activity Type	Water		Air		Habitat		Wildlife		
	Local	Regional	Local	Regional	Aquatic	Terrestrial	Birds	Mammals	Other
Construction:									
Excavation									
Blasting									
Truck traffic									
Foot traffic									
Temporary roads									
Other:									
Operation:									
Truck traffic									
Filling activities									
Leachate collection									
Gas collection									
Other:									
etc...									

The matrix cells can then be filled with information about direction and magnitude of impact. A common approach is to split the cell diagonally, placing a "+" or "-" sign in the upper half to indicate positive or negative direction of impact (or a "0" for no impact), and some numerical score or ranking in the lower half. The finished cell entry might look something like this, where "7.5" might be a score out of ten, indicating that this action is likely to have a fairly serious negative impact on this environmental component.

182

ily general and simplified, if completed for a detailed list of project activities and environmental components, it can provide an excellent overview of probable areas and severity of impact.

Example 2: Summary impact matrix

Another commonly used matrix type is one which summarizes all impacts of the project on a detailed list of environmental components. Because this matrix type reduces the amount of detail shown for individual project activities, it allows more detail to be shown about whether the expected impacts are long- or short-term, reversible or irreversible, and so on. An example might look like Figure 15.4.

Figure 15.4: Example summary impact matrix

Environmental component	Adverse impact	Beneficial impact
Freshwater aquatic ecosystems		
Marine ecosystems		
Fisheries		
Wetlands		
Estuaries		
Rare and endangered species	*In this case, the body of the matrix would be filled with information about whether expected impacts were short-term or long-term; reversible or irreversible; extreme, moderate, or negligible. Impacts would be listed under the appropriate column: adverse or beneficial. The format gives enough space that phrases regarding the nature of the impact (e.g., "increased nitrate") can be added in addition to acronyms and numerical scores.*	
Mammals		
Birds		
Terrestrial habitats		
Forests		
Surface waters		
Groundwater		
Soil physical and chemical properties		
Human settlements and cultures		
Viability of local/regional businesses		
Property values		
Human health		
Tourism and recreation		
Aesthetic, religious and historical sites		
etc...		

Costing the impacts

Assigning costs to intangible factors like habitat disruption is difficult and controversial, but not impossible. Case Study 9 on Costa Rica describes some possible techniques. Typical approaches include surveys to determine a community's "willingness to pay" for a given attribute; contingent valuation analysis (see, for example, Roberts et al. 1991); hedonic price analysis, which examines the difference in property values between sites that have and do not have a given attribute such as low noise levels; qualitative ranking systems; and the travel-cost analysis technique described in Case Study 9. Sometimes, interviews with local experts—"key informants"—can be helpful in deciding the value of a given enviromental component, and thus deducing the cost of an impact. A central consideration in this type of analysis is that the cost estimates developed should represent the views of the community, not just the analyst. Literature values are of little help here, because one community, or even one neighborhood, will place a higher or lower value on a given attribute than a similar community or neighborhood elsewhere.

Sum all costs

When we have determined the capital costs, the operating and maintenance costs, any additional costs for specialized treatment or disposal, labor and energy costs (as far as is possible), social costs, and environmental costs, we can arrive at a total, and presumably at least more complete, "cost" for the landfill. That total cost can then be allocated on a per-ton or tonne basis for the expected mass of waste that the landfill will accommodate.

In Hirshfeld et al.'s 1992 analysis, which excludes consideration of a broader range of environmental impacts, the authors arrived at the following revised estimates of the possible "cost" of the Durham landfill project. Recall that the city had estimated a tipping fee of $35.20/tonne based on conventional capital and operating costs. Hirshfeld and his colleagues estimated the following additional costs (note that they considered opportunity costs, which are not described in the foregoing text):

Tipping fee	$ 35.20/tonne
Leachate, gas and monitoring costs	24.20/tonne
Property depreciation	1.10/tonne
Lost possible revenues ("opportunity costs") for landfill site	10.90/tonne
Opportunity costs for adjacent properties	0.40/tonne
Total "cost"	**$71.80/tonne**

Hirshfeld and his colleagues therefore estimate a true cost for the landfill (as reflected in the tipping fee) that is more than double that calculated by the city. If we were to include additional "costs" of environmental impacts like habitat disruption, and air and water contamination (even if these are unlikely given the proposed level of capture and treatment), the true cost of the project would probably be higher still.

The lessons to be drawn from this analysis are several. First, it is clear that the "costs" of a project with environmental implications can be calculated in different ways. It is also clear that conventional calculation approaches probably significantly underestimate the true "costs" of landfills. And finally, it is likely that these underestimated costs, however inaccurate, can play an important role in public decision-making, in this case potential approval of a project whose impacts may not have been adequately thought through. At the very least, the impacts of this project may not be fully apparent to the community simply because of the simplified approach taken in estimating tipping fees.

5. *How Can I Use My Findings to Reach a Solution?*

Siting a landfill is a very common problem throughout much of the world. This case presents a North Carolina example that is nevertheless typical of most other landfill projects elsewhere.

1. What is the problem?

In Section 2, we identified the problem as how best to assess the full range of social, economic, and environmental impacts of the proposed Durham landfill project.

2. In what ways do human activities have impact on the natural environment to cause "a problem"? How do these mechanisms give you clues to possible solutions?

Developed countries produce a lot of waste. In some areas, nonhazardous wastes are incinerated, but in many others landfilling is the disposal option that has proved most feasible. Landfilling impacts the environment in a variety of ways (see Box 15.1), including water, air, and noise pollution, diminished aesthetic quality, and reduced property values. Tipping fees are usually based only on the capital costs and revenues of a landfill, but this range of impacts suggests that the true costs of such a project may be much greater than are normally estimated. This leads us to extend routine cost analysis to include a wider range of activities and to do so in such a way that the analysis reflects the opinions and values of a range of interested parties.

3. What governments are responsible for the issue? Whose laws may apply?

As in Case Study 13 (Coors Brewing), the U.S. (federal) Environmental Protection Agency has overriding authority for waste management activities, although some of the implementation of EPA's rules is delegated to the state governments, often through enabling state law; both federal and state governments would therefore be involved in this case. Municipal governments have a regulatory role in siting landfills because of their land-use-planning responsibilities. In this case, the Durham city government will also operate the landfill.

4. Who has a stake in the problem? Who should be involved in making decisions?

Stakeholders in this case include landfill operators, waste haulers, waste generators (including Durham residents). The governments listed in step 3 would have a clear interest, particularly the city of Durham. Neighboring landowners would be affected through property value decreases, noise, dust, and similar impacts.

5. In the view of your decision-making group, what are the attributes of a satisfactory solution? In other words, when will you be satisfied that the problem is "solved"?

A satisfactory solution here is one which yields a comprehensive and accurate estimate of true total costs for the project. As in Case Study 9, the interpretation of "comprehensive" and "accurate" will vary depending on the decision-making group.

6. How will you evaluate (test, compare) potential solutions?

Section 4 provides information on basic cost estimation approaches and more complex assessment of environmental impacts using two matrix techniques. These methods will enable you to compare several different landfill sites in a quantitative fashion. The choice of an analytical method is essentially a subjective one, usually made by the decision-making group.

7. What are all the feasible solutions to the problem?

As in Case Study 9, there are many ways to configure a cost analysis, and indeed this is one of the central points of debate in environmental economics. Feasible analytical approaches will be those that reflect the values and priorities of the community, especially around valued landmarks, activities, and scenery. These are elements that can only be decided at the community level.

8. Which solutions work "best" in terms of the attributes you identified in (5)?

Ultimately, we need this kind of analysis to decide whether one site is better than another, to set tipping fees, and to devise compensation programs for those affected by the project. A "best" solution will therefore be one that reflects consensus among the stakeholders on these various end points. Analyses that fail to secure community support are, for that community, inadequate.

9. Which solution will be easiest to implement?

There is no question that the traditional cost-estimation approach is much faster, cheaper, and generally simpler than the more expanded analysis described in this case study. If the traditional approach (neglecting, for example, property values and environmental impacts) meets the community's needs, it would certainly be the easiest to implement. More commonly, however, the analyst encounters objections to the scope of the analysis, ultimately forcing a more comprehensive approach. The analyst can save time and hard feelings by consulting stakeholders on the scope of the analysis before proceeding with the analysis itself.

10. What steps are needed for successful implementation? Who will pay? Who will monitor progress?

A simple, traditional analysis requires nothing more than estimates of the kind presented in Boxes 15.2 and 15.3. More complex analysis of property values and environmental impacts may require considerable field work to survey landowners and experts, compile literature research, and analyze this information using quantitative or qualitative techniques. Such work will require detailed time and cost projections, individual and agency responsibilities, and so on. In either case, but more importantly in the latter, data custody and reporting responsibilities should be outlined clearly in the implementation plan.

6. Where Can I Learn More About the Ecosystem, People, and Culture of Communities Using Landfilling?

The literature contains a wide range of materials and case studies on landfill site selection, various technical aspects of landfill design and pollution control, and the economics of waste production and disposal. Here are some sources that may be useful in learning about this field.

C. J. Booth and P. J. Vagt. 1990. Hydrogeology and historical assessment of a classic sequential-land use landfill site, Illinois, USA. *Environmental Geology and Water Sciences* 15(3): 165-178.

I. Frantzis. 1993. Methodology for municipal landfill sites selection. *Waste Management and Research* 11(5): 441-451.

J. Glenn and D. Riggle. 1991. The state of garbage in America. *Biocycle* 32(4): 34-38.

S. Hirshfeld, P. A. Vesilind, and E. I. Pas. 1992. Assessing the true cost of landfills. *Waste Management and Research* 10(6): 471-484.

S. D. Minor and T. L. Jacobs. 1994. Optimal land allocation for solid- and hazardous-waste landfill siting. *Journal of Environmental Engineering* 120(5): 1095-1108.

A. C. Nelson, J. Genereux, and M. Genereux. 1992. Price effects of landfills on house values. *Land Economics* 68(4): 359-365.

R. K. Roberts, P. V. Douglas, and W. M. Park. 1991. Estimating external costs of municipal landfill siting through contingent valuation analysis: a case study. *Southern Journal of Agricultural Economics* 23(2): 155-164.

H. W. Sheffer, E. C. Baker, and G. C. Evans. 1971. Case studies of municipal waste disposal systems. U.S. Department of the Interior, Bureau of Mines, Information Circular 8498.

J. M. Suflita, C. P. Gerba, R. K. Ham, A. C. Palmisano, W. L. Rathje, and J. A. Robinson. 1992. The world's largest landfill: a multidisciplinary investigation. *Environmental Science and Technology* 26(8): 1486-1494.

"How can we approach the problem of 'scoping' an environmental impact assessment?"

Toronto area

1. What Is the Background?

A lakefilling project is proposed

Like many large municipalities in Ontario, Metropolitan Toronto is located on the shore of a major water body, Lake Ontario. Since Toronto's earliest days, the lake has provided a means of transportation and communication with other urban centres. As the city grew, marshy areas of the shoreline were filled in, often using excavated material from building sites elsewhere in the city. Now, the shoreline bears little resemblance to the prehistoric condition.

Recently, the local Conservation Authority (a watershed management agency) developed a proposal to create a new parkland/marina complex by this lakefilling technique. They have identified a suitable site on a strip of undeveloped shoreline west of the city. They argue that the development is necessary because it will provide recreational parkland and small boat moorings in the urban environment.

The site exhibits good environmental quality

The site is flanked by residential development to the east and west. On the north end of the property, a psychiatric hospital sits on spacious, neatly landscaped grounds. A major east-west street is located just north of the hospital, with a long winding driveway leading from the street into the hospital grounds.

Water quality in the lake near the shoreline is good, with occasional high levels of bacteria, especially after thunderstorms. The bacteria come from storm sewer discharges (draining road and parking lot surfaces) and combined sewer overflows (overflows of sewers carrying mixed stormwater and sanitary sewage). Sewer pipes of both kinds discharge into the lake near the proposed site. At present, there is a strong west-to-east current along the shoreline, which quickly disperses these sewage flows, although some chemicals present in the sewage may bioaccumulate in aquatic plant and animal tissue. Prevailing winds are also from west to east for much of the year.

Table 16.1: Typical water quality results, Lake Ontario nearshore zone near Toronto	
Total phosphorus	0.03 mg/L
Ammonia	0.04 mg/L
Conductivity	370 μmho/cm²
Lead	0.4 mg/L
Copper	0.01 mg/L
Zinc	0.01 mg/L
Mercury	0.02 μg/L
Dissolved oxygen	5.02 mg/L
Fecal coliform bacteria	<10 per 100 mL

The bottom substrate (sediment) in the lake is generally sand and gravel, providing good potential for fish spawning. Until recently, whitefish (now absent from the area) were known to spawn along this stretch of shoreline, and other large fish species still enjoy the granular sediments and fast-flowing water.

Under the proposed lakefilling project, thousands of truckloads of excavated soil would be dumped into the lake and gradually built out into a C-shaped peninsula. The slopes of the lakefill would be protected from wave action by "armoring" with large stones and boulders. The mouth of the "C" will face westward, with the interior basin constructed as a marina with space for 400 pleasure craft. The remainder of the lakefill would be used as parkland, with picnic tables, concrete paths for strolling, and similar amenities.   Access to the site would be provided by an extension of the existing driveway down to the waterfront and onto the lakefill surface. Maps showing the present site configuration and the proposed development are given in Figure 16.1. Water quality information appears in Table 16.1.

An environmental assessment is required

Assume that you are employed by a local consulting firm hired by the Conservation Authority to conduct the environmental assessment. A lot will hinge on your work: if the provincial government finds your work to be incomplete or inaccurate, they may ask your company to do it again. Or they may approve it but require that it proceed to public hearings. If you can't defend your assessment adequately at the hearing, your job and your personal reputation may be in jeopardy. How should you begin?

2. What Problem Are We Trying to Solve?

Ultimately, we want to determine the most significant environmental impacts of the project. But the first problem we have to solve is what to include in that analysis.

A good place to start would be to find out what the law *requires* you to do. In Ontario and most other provinces and states, the "rules" are clearly laid out in a statute (law) or regulation (attachment to a law laying out specific requirements). The requirements of the Ontario Environmental Assessment Act are similar to those in most other jurisdictions, and are more stringent than some. They provide a useful model for developing an environmental assessment for the proposed Toronto lakefilling project.

(There are a number of considerations that are not explicitly addressed in the Environmental Assessment Act. Among them are the feelings and concerns of local residents. One of the major challenges in conducting a good environmental assessment is therefore to arrive at a

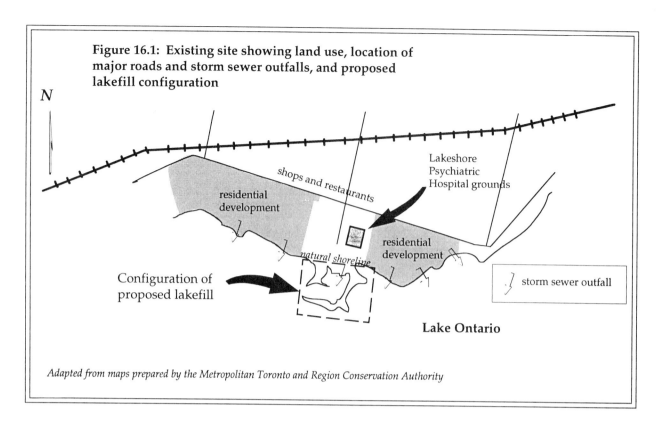

Figure 16.1: Existing site showing land use, location of major roads and storm sewer outfalls, and proposed lakefill configuration

Adapted from maps prepared by the Metropolitan Toronto and Region Conservation Authority

working definition of "the environment" that will potentially be affected by the project, and to develop a consultation process that allows you to gather input from those who will be affected by it. In this case, local residents were deeply concerned about potential changes in their quality of life, particularly their peaceful enjoyment of lakefront views and wildlife, and possible impacts on the value of their properties. The environmental assessment hearing allowed these people a voice, but their concerns were probably not adequately represented in the assessment document that was prepared by the project's consultants. This case is not expressly concerned with consultation processes, but rather with developing a "scope" for the assessment. (Case Study 4, from New Zealand, provides an example of public consultation mechanism in a different kind of land-use-planning situation.)

Looking over the requirements of the Ontario Environmental Assessment Act is a pretty daunting task. It seems like just about anything should be part of your assessment. Yet most projects don't have unlimited time and money to spend on analysis, so the analyst will somehow have to limit the components of the assessment to a manageable level. This process is called "scoping" and is described in more detail in the next section. To develop an adequate scope for your assessment, you need to ask two questions:

1.   What time, space, and population boundaries should I put on the assessment?

and

2.   What are the most important issues likely to be?

We will discuss approaches to answering these questions in the next sections.

Box 16.1: Requirements under Ontario's Environmental Assessment Act

In essence, the act requires that the person proposing an undertaking ("the proponent") must, before proceeding, prepare an environmental assessment describing the undertaking and the alternatives to it, all of their effects on the environment, and proposed measures to mitigate these effects. The **"environment"** *is defined as:*

- *physical features (air, land, water)*

- *any building, structure, machine, or other device or thing made by humans*

- *any solid, liquid, gas, odor, heat, sound, vibration, or radiation resulting directly or indirectly from the activities of humans*

- *biological subjects (plant and animal life, including humans)*

- *human and ecological systems (the social, economic and cultural conditions that influence the life of humans or a community)*

- *any part or combination of the foregoing*

- *the interrelationships between any two or more of them*

An **"undertaking"** *is defined as:*

- *any enterprise or activity, or even a proposal, plan, or program for an enterprise or activity by or on behalf of the province, a public body, or a municipality*

(Public projects are included unless exempted; private projects are excluded unless and until designated by regulation.)

The assessment must include

- *a description of the purpose of and reason for the undertaking*

- *descriptions of and rationales for the alternative methods of carrying out the undertaking (for example, different construction approaches, or phasing-in of construction) and the alternatives to the undertaking (for example, the "do nothing" option, or different ways of achieving the same end (in this case, additional recreational opportunities in the urban setting)*

And it must describe

- *the environment that will be affected directly or indirectly*

- *the effects that will be caused to the environment*

- *the actions necessary to prevent, change, mitigate or remedy the effects*

Finally, it must include

- *evaluations of the advantages and disadvantages to the environment of the undertaking, of the alternative methods of carrying out the undertaking, and of the alternatives to the undertaking*

At minimum, there is therefore an information-gathering exercise that forces proponents to consider factors they otherwise probably would not consider, or would consider only in a vague and undocumented way.

3. *What Components of the Environment Are Affected, and How?*

Three "environments"

There are three environments that we are concerned with in "scoping" this environmental assessment. First, there is the biophysical environment of the nearshore zone and adjacent shoreline areas. This "environment" includes the plants and animals living there and the abiotic (nonliving) parts of the environment like the water, sediment, terrestrial soils, air, and the various substances flowing through that system. Some of this environment is discussed in Section 2; in general, it is a typical temperate aquatic environment with good water and air quality, good species diversity, and low levels of most contaminants.

There is also the "built" environment of the psychiatric hospital and (if constructed) the marina facilities, parking lots, and other recreational facilities. A short distance from the site are other parts of the environment we may be concerned with—things like local office buildings and homes. This "built" environment also includes elements like storm sewer outlets, that discharge storm runoff from streets and parking lots into the lake.

Finally, there is the human environment—the people that live nearby, the patients at the psychiatric hospital, the potential users of the new facility, and so on. This environment includes not only the people themselves (who in a sense are a part of the natural environment), but also their culture, their quality of life, their economic systems, their values, their wishes, and their fears.

Many concepts of those environments

The most challenging aspect of scoping an environmental assessment is that there are many views of what is, and is not, important in the natural, the built, and the human environment. For the purposes of this case study, it is probably less important to understand exactly which species of snake or butterfly live in the study area than to understand the larger issues that people worry about. As the scoping proceeds, certain key issues should begin to emerge as worthy of particular attention (see Box 16.4). When these have been identified, detailed planning can proceed with special attention to those areas.

Section 4 discusses some ways of approaching the problem of scoping, including the essential aspect of stakeholder consultation.

TIP *You must be ready to justify any decisions you make about narrowing your scope. For example, if you decide to ignore any impacts on insects, you must be ready to explain why that is a sound decision. You might have made that decision because no rare insects are known to be present in the area, or because insects are not likely to be affected by the impacts you predict, or a similar reason.*

4. How Can I Analyze This Information?

Set boundaries on space, time, and population

You can begin to answer this question by breaking down the two questions at the end of Section 2. Let's take the problem of *boundaries* to begin with. What is meant by the term "boundaries"? It's really the limits that you choose to set on your study. Note the word "choose": you won't find any books to tell you what belongs in your analysis and what should be excluded, because those decisions will be different for every different situation. You can, however, break the problem of boundaries down into specific components. Here are some example questions:

What geographic area is likely to experience direct and immediate impacts from the proposed project?

What additional area may experience secondary or indirect effects?

In many cases, you will find that the immediate impact of the development is limited to a 5-km radius of the site. You will want to focus much of your analytical effort on that area. A wider area may be affected indirectly, for example by traffic passing through it. You will want to consider impacts on this area too, but perhaps in less detail.

Over what time period, or periods, should impacts be assessed?

Many major projects will involve construction activities and the use of heavy machinery. This construction phase will have its own special impacts and concerns (for example dust, noise, and vibration) that are separate and different from those that will occur during system operation or closeout. You will need to understand how long such a construction phase might last and the nature of the impacts that might occur in it.

A different set of concerns is likely to arise during the operational phase, the time following completion of construction when the system or device is operating normally. The operational phase is usually much longer than the construction phase—perhaps 25 years as compared to 3 or 4 years for construction.

Finally, you will need to understand the possible implications of "closeout" or decommissioning of the project. When the operational life of a structure or facility is finished, some action needs to be taken to close it down and seal it off from the environment. In some cases, this will mean dismantling it; in others, it may mean capping an underground structure. Whatever the necessary action, you will need to understand what is involved with closeout, how long the closing down will take, and whether an ongoing care is required. If perpetual care is needed, as it would be in the case of a landfill or abandoned mine, the post-closeout phase, with its associated impacts, could continue for tens or hundreds of years.

Box 16.3: Choosing a "focus"

How do you decide on an appropriate focus for your analysis?

The best approach is probably a two-stage process. The first step is to decide on some reasonable limits—boundaries—for your study. This will get you thinking about a particular geographic area and time period, and help remind you what populations of humans, other animals, and plants might be affected by the project.

Now think about the likely nature of impacts on those populations. In some cases, economic impacts, such as property values, sales revenues from commercial operations, and so on, will be critical. In other areas, damage to "heritage" properties, such as graveyards or historic sites, could be a major concern. In still others, impacts on human health may be an important issue.

Finally, decide which of these impacts is most important. Different people will have different ideas about what is important and what is not. Try to determine what the majority view is. If you have covered a good range of viewpoints in your key informant interviews or other research, two or three main issues will likely emerge.

When you've gone through the process of focusing your analysis, you are in a good position to defend your final study design. If challenges are raised, for example in a public hearing, you can reply that you have consulted x sources, of which y were of the opinion that these were the key issues. You may even be in a position to rebut the challenge by referring to a key informant interview with the challenger!

> "You may remember, Ms. Smith, that when we spoke on June 25, you agreed that consideration of flying squirrel habitat is probably secondary to concerns about human health in this project."

What plant and animal populations should be considered in assessing "impact"?

There are potentially thousands of plant and animal species in the area you are assessing. Remember that insect, amphibian, reptile, and bird species may be just as much of concern as furbearing mammals. Similarly, your area may support rare species or subspecies of plants that should be protected from impact. Finally, humans are likely to be among the populations of animals inhabiting your area. You may wish to consider subgroups of the human population separately, such as children, women of reproductive age, adult men, elderly or infirm residents, and so on.

Your choice of which populations should be considered could begin with general considerations. For example, you could find out whether any endangered plants or animals are known to exist in the study area. You could determine whether any domestic animals might be affected, whether pets such as dogs and cats, or domestic animals such as pigs, cows, horses, sheep, and chickens.

Now let's look at the problem of *focusing*. In the early days of environmental assessment, people thought that a good analysis was one that included a paragraph on every possible plant and animal, at every possible time and place, that could conceivably be impacted by the proposed project.

Now we know better. An environmental assessment is most useful when it gives an overview of the situation, but focuses on a small number of issues that are generally agreed to be important (see Box 16.3). If you have conducted key informant surveys (see Box 16.4), these issues will have emerged early in the process: almost everyone you interview will mention them in one way or another.

Where can you find the information you need?

Even when you've set some pretty strict boundaries around the problem, you may still be overwhelmed by the size of your task. Don't be discouraged!

If you've set your boundaries thoughtfully, they should give you clues to literature on specific aspects of the problem. For example, you could go to the library and look up information on soil types, climate, or prevailing winds for the specific area you are considering. Population boundaries help you narrow your search to information about a particular species of bird, or medical literature on health effects of a given toxin on pregnant women.

You will also need to find people who are knowledgeable about the issue: your key informants. To begin your search, you may need only one name. A successful interview with that individual can lead to further contacts, who in turn will lead to other names. You can find that first name in several ways. In a city, you could contact members of the city's Works or Planning Department. You could also visit a local consulting engineer or call a local residents' association or public interest group. Newspaper articles, especially in local papers, can give clues about organizations and individuals who are interested in your particular issue or related projects.

(Finding the information you need to assess the actual impacts is a separate problem. It is likely that some data will exist for aspects you want to consider; but much data also will be missing. You may have to rely on other researchers' studies for some answers. Or you may be able to simulate aspects of the environment before and after disturbance using computer simulation models.

Clearly, there's no "right" answer in scoping an environmental assessment. The difficulty an environmental manager faces is that the law requires you to consider impacts on all possible aspects of the environment, but your time and money to do this are limited.

Box 16.4: The key informant interview

Let's face it: few of us have all the answers. This is especially true when it comes to deciding what is important or unimportant in resolving an environmental issue.

Nowhere is this more evident than in "scoping" an environmental assessment. Your final assessment will be subject to extensive scrutiny, not all of it from sympathetic perspectives. It's wise to anticipate all the issues early on so that you can address—or dismiss—them in your report.

*One excellent way of finding out what **might** be important is to ask knowledgeable individuals from a variety of backgrounds. This type of interview is often referred to as a "key informant interview." To be effective, such interviews should include people from all sides of the issue (in this case, people who are likely to support the project and those who are opposed to it). You could include government representatives, public interest group staff members, local residents' association leaders, and professional staff from local industries or consulting operations.*

In the interview, you should not be afraid to admit that you are looking for new ideas and viewpoints and that you would welcome the names of additional contacts. Then simply be frank with your source and ask what he or she considers the key issues in the situation—what should be included in terms of space, time and population, and so on.

The more people you interview, the broader your understanding of the issues will be. And the better your understanding, the more persuasive your arguments for including or excluding a given component of the analysis.

The "best" solution is one that is thorough and well-justified and has the concurrence of a wide range of stakeholders. It will not be arbitrary ("I don't like snakes so I left them out.") and it will avoid obvious bias ("Everybody knows that this project will be good for the town's economy.") It will represent the views of many people fairly and objectively, and it will defend its scope on the basis of objective research or broad consensus. The research that you use may be biology, sociology, economics, law, physics, chemistry, or another discipline.

The true test of your scoping efforts will be the reception they receive from government reviewers and interested members of the public. If you have done your consultation well, you should find no surprises in these responses. If, on the other hand, you have designed an assessment that reflects primarily your own views, or an incomplete research effort, you will almost certainly find yourself faced with redoing the work.

As this discussion suggests, the more specific you can be about your boundaries, the more straightforward your research task. You may also find that it's easier to formulate clear questions for interviews if you can be very specific about what you want to know. For example, it's better to ask, "To your knowledge, are there any populations of the Kirtland's warbler within the area shown on this map?" than it is to ask, "Do you know of any rare bird species around here?"

5. How Can I Use My Findings to Reach a Solution?

"Scoping" an environmental assessment is a useful background for many research activities, where the analyst must decide what to include and what to ignore. The lakefilling case is not a particularly common one but it is certainly a typical "scoping" problem.

1. *What is the problem?*

In Section 2, we identified the problem as "scoping" an environmental assessment for a Lake Ontario lakefilling project, including setting boundaries on the time, space and populations that will form part of the analysis, and choosing key issues as a focus for the work.

2. *In what ways do human activities have impact on the natural environment to cause "a problem"? How do these mechanisms give you clues to possible solutions?*

This case describes a relatively pristine stretch of lakeshore near a large city. There is residential and institutional development nearby and commercial development a little further away. The lakefill will alter this environment dramatically, by introducing immense quantities of excavated fill into the nearshore zone, thus blocking longshore currents, altering fish spawning habitat, reducing the dispersion of stormwater and combined sewer overflow discharges entering the lake, and possibly altering property values and quality of life for nearby residents. The size of the development and the large quantity of fill to be used in it suggest that we will need to predict impacts over many years, though a multiyear construction phase, a multidecade operation phase, and possibly a decommissioning phase 50 or 100 years into the future. The most acute impacts will be felt in the immediate vicinity of the project, probably within a radius of 0.5 km. But we may well want to examine impacts over major transportation corridors used by the trucks bringing fill to the project. The human population, particularly local residents, psychiatric patients, and potential marina users, are likely to be more affected than any wildlife in the area. Fish populations may be affected by increased sedimentation from filling activities and reduced current velocities.

3. *What governments are responsible for the issue? Whose laws may apply?*

The Ontario provincial government probably has the lead role here, through the Ontario Environmental Protection Act. The municipal government (the local municipality of Etobicoke and/or the regional municipality of Metropolitan Toronto) would also have interest in land-use planning and traffic control. The Conservation Authority, a branch of the provincial government, is in fact the proponent in this project..

4. *Who has a stake in the problem? Who should be involved in making decisions?*

Likely stakeholders include the Conservation Authority, the provincial Ministry of the Environment and Energy, the local and regional municipalities, local residents, future users of the marina and other recreational facilities, and environmental nongovernment organizations concerned with protecting the lakeshore environment. Local businesses and the psychiatric hospital would also be important stakeholders here.

5.   In the view of your decision-making group, what are the attributes of a satisfactory solution? In other words, when will you be satisfied that the problem is "solved"?

The scope will be adequate when it has the approval of all major stakeholders. The project cannot, in fact, go forward unless the scope of the environmental assessment has been approved by the Ministry of the Environment and Energy in consultation with other provincial agencies.

6.   How will you evaluate (test, compare) potential solutions?

There is no need for quantitative evaluation of alternative scoping approaches in this case. The best way to evaluate possible approaches is to discuss them with the key stakeholders, receive feedback, and refine the scope to address any concerns that have been expressed. Section 4 discusses the need for stakeholder consultation in scoping. Although consultation is not required under the law, experience has shown that analysts who have consulted widely, and thus have adequately scoped their assessments, encounter far fewer objections and delays in the approval process than those who fail to consult.

7.   What are all the feasible solutions to the problem?

There are many possible approaches to scoping a problem like this. Consultation with key informants is especially important in identifying and remedying weaknesses in the plan. The process therefore becomes one of iterative improvement rather than objective comparison.

8.   Which solutions work "best" in terms of the attributes you identified in (5)?

The "best" scope is one that reflects community values and priorities and that has the support of the major stakeholders in the issue. The details of the scope, and the focus, will vary depending on the particular place and time, and the individual personalities and interests involved. Role play can provide a useful mechanism for students to gain insight into the dynamics of scoping and the progressive development of a consensus-based scope.

9.   Which solution will be easiest to implement?

The Environmental Protection Act was developed expressly to force objective comparison of alternative approaches and avoid the obstacles that can arise in public debate about public projects. Ease of implementation will depend on the adequacy of the scope, and thus indirectly on the adequacy of the pre-submission consultation. If you have consulted widely and understand the full range of issues, and your boundaries and focus reflect the values of key stakeholders, you should be allowed to proceed to a full environmental assessment.

10.   What steps are needed for successful implementation? Who will pay? Who will monitor progress?

Part of scoping is determining the activities that will form the actual environmental assessment activity: what data will be used, over what time period, and analyzed with which methods, for example. The scope for an EA is essentially an implementation plan, identifying the timing and cost of specific activities and laying out the agency and individual responsibilities for data collection, analysis, and reporting.

5. *Where Can I Learn More About Environmental Assessment?*

There is a large literature on environmental impact assessment. The following give an introduction to that literature.

Floor Brouwer. 1987. *Integrated Environmental Modelling: Design and Tools.* Kluwer, Boston.

Paul N. Cheremisinoff. 1977. *Environmental Assessment and Impact Statement Handbook.* Ann Arbor Science, Ann Arbor.

Derek V. Ellis. 1989. *Environments at Risk: Case Histories of Impact Assessment.* Springer-Verlag, New York.

David James. 1994. *The Application of Economic Techniques in Environmental Impact Assessment.* Kluwer, Boston.

J. G. Rau and D. C. Wooten. 1980. *Environmental Impact Analysis Handbook.* McGraw-Hill, New York.

Index